Principles of Geology

Principles of Geology

Leslie Taylor

R CALLISTO REFERENCE

www.callistoreference.com

Callisto Reference,
118-35 Queens Blvd., Suite 400,
Forest Hills, NY 11375, USA

Visit us on the World Wide Web at:
www.callistoreference.com

ISBN: 978-1-64116-580-8 (Hardback)

Cataloging-in-Publication Data

Principles of geology / Leslie Taylor.
    p. cm.
Includes bibliographical references and index.
ISBN 978-1-64116-580-8
1. Geology. 2. Earth sciences. I. Taylor, Leslie.
QE26.3 .P75 2022
550--dc23

Table of Contents

Preface

The earth science which deals with the solid earth, the rocks of which the solid earth is made up of, and the processes which are involved in their change is called geology. Along with Earth, geology also deals with the solid features of other planets and natural satellites. There are various methods which are used in geology such as geophysical techniques, numerical modeling, rock description and chemical analysis. Some of the practical applications of this field are providing insights into past climate change, evaluating water resources, remediating environmental problems, and mineral and hydrocarbon exploration. Geology is also involved in the dating of different rocks and fossils. The methods which are used for dating are broadly divided into two types, namely, relative dating and absolute dating. This book discusses the fundamentals as well as modern approaches of geology. It consists of contributions made by international experts. The book will serve as a valuable source of reference for graduate and post graduate students.

A foreword of all Chapters of the book is provided below:

Chapter 1 - The branch of Earth science which deals with solid Earth, rocks and the processes involved in their change is called geology. It also studies the solid features of extraterrestrial planets and natural satellites. The topics elaborated in this chapter will help in gaining a better perspective about the geology.; **Chapter 2** - The branch of natural science which focuses on the physical constitution of the Earth and its atmosphere is called geoscience. There are a number of sub-disciplines within geoscience such as geobiology, geochemistry and geophysics. This chapter discusses in detail these sub-disciplines of geoscience.; **Chapter 3** - The scientific theory which describes the motion of the different plates of Earth's lithosphere is called plate tectonics. Some of the areas studied within plate tectonics are tectonic landforms, supercontinents, earthquakes and volcanoes. The diverse aspects of plate tectonics as well as these focus areas have been thoroughly discussed in this chapter.; **Chapter 4** - The branch of geology which studies the crystal structure, chemistry and physical characteristics of minerals and mineralized artifacts is termed as mineralogy. Petrology refers to the branch geology which deals with the study of rocks and the conditions under which they form. The topics elaborated in this chapter will help in gaining a better perspective about the diverse aspects of mineralogy and petrology.; **Chapter 5** - The study of sediments like sand, clay and silt along with the processes which result in their formation are studied within the discipline of sedimentology. This chapter closely examines the key aspects of these sediments to provide an extensive understanding of sedimentology.; **Chapter 6** - The domain of geology which studies the movement and distribution of groundwater in the soil and rocks of the Earth's crust is called hydrogeology. The key areas of study within this discipline include aquifers and the hydrological cycle. This chapter has been carefully written to provide an easy understanding of these facets of hydrogeology.

I would like to thank the entire editorial team who made sincere efforts for this book and my family who supported me in my efforts of working on this book. I take this opportunity to thank all those who have been a guiding force throughout my life.

Leslie Taylor

Chapter 1

Introduction to Geology

The branch of Earth science which deals with solid Earth, rocks and the processes involved in their change is called geology. It also studies the solid features of extraterrestrial planets and natural satellites. The topics elaborated in this chapter will help in gaining a better perspective about the geology.

Geology is the science and study of the Earth, its composition, structure, physical properties, history, and the processes that shape it. It is one of the Earth sciences. In order for mankind to live in harmony with our environment we have to understand it. In this context knowledge of geology is an essential component.

Geologists have helped establish the age of the Earth at about 4.5 billion years and that the Earth's interior is a series of concentric layers of different kinds of materials:

- The iron-rich core,

- A thick rocky shell, the mantle with its outermost layer being the asthenosphere,

- The lithosphere, which includes the exposed surface of Earth's crust.

The crust is fragmented into tectonic plates that move over the rheic asthenosphere via processes that are referred to collectively as plate tectonics. Geologists employ an unusual quality of imagination in visualizing processes such as the movement of tectonic plates taking place over millions and hundreds of millions of years. Their work in establishing strong evidence for an ancient origin of the earth has been a key factor supporting conventional scientific models in the ongoing skirmishing between science and religion.

Geologists help locate and manage the Earth's natural resources, such as petroleum and coal, as well as metals such as iron, copper, and uranium. Additional economic interests include gemstones and many minerals such as asbestos, perlite, mica, phosphates, zeolites, clay, pumice, quartz, and silica, as well as elements such as sulfur, chlorine, and helium.

Astrogeology refers to the application of geologic principles to other bodies of the solar system. However, specialized terms such as *selenology* (studies of the Moon), *areology* (of Mars), are also in use.

Important Principles of Geology

There are a number of important principles in geology. Many of these involve the ability to provide the relative ages of strata or the manner in which they were formed.

- The Principle of Intrusive Relationships concerns crosscutting intrusions. In geology, when an igneous intrusion cuts across a formation of sedimentary rock, it can be determined that the igneous intrusion is younger than the sedimentary rock. There are a number of different types of intrusions, including stocks, laccoliths, batholiths, sills, and dikes.

- The Principle of Cross-cutting Relationships pertains to the formation of faults and the age of the sequences through which they cut. Faults are younger than the rocks they cut; accordingly, if a fault is found that penetrates some formations but not those on top of it, then the formations that were cut are older than the fault, and the ones that are not cut must be younger than the fault. Finding the key bed in these situations may help determine whether the fault is a normal fault or a thrust fault.

- The Principle of Inclusions and Components states that with sedimentary rocks, if inclusions (or *clasts*) are found in a formation, then the inclusions must be older than the formation that contains them. For example, in sedimentary rocks, it is common for gravel from an older formation to be ripped up and included in a newer layer. A similar situation with igneous rocks occurs when xenoliths are found. These foreign bodies are picked up as magma or lava flows, and are incorporated later to cool in the matrix. As a result, xenoliths are older than the rock which contains them.

- The Principle of Uniformitarianism states that the geologic processes observed in operation that modify the Earth's crust at present have worked in much the same way over geologic time. A fundamental principle of geology advanced by the eighteenth-century Scottish physician and geologist James Hutton is that "The Present is the Key to the Past." In Hutton's words: "the past history of our globe must be explained by what can be seen to be happening now."

- The Principle of Original Horizontality states the deposition of sediments occurs as essentially horizontal beds. Observation of modern marine and nonmarine sediments in a wide variety of environments supports this generalization (although cross-bedding is inclined, the overall orientation of cross-bedded units is horizontal).

- The Principle of Superposition states a sedimentary rock layer in a tectonically undisturbed sequence is younger than the one beneath it and older than the one above it. Logically a younger layer cannot slip beneath a layer previously deposited. This principle allows sedimentary layers to be viewed as a form of vertical time line, a partial or complete record of the time elapsed from deposition of the lowest layer to deposition of the highest bed.

- The Principle of Faunal Succession is based on the appearance of fossils in sedimentary rocks. As organisms exist at the same time period throughout the world, their presence or (sometimes) absence may be used to provide a relative age of the formations in which they are found. Based on principles laid out by William Smith almost a hundred years before the publication of Charles Darwin's theory of evolution, the principles of succession were developed independently of evolutionary thought. The principle becomes quite complex, however, given the uncertainties of fossilization, the localization of fossil types due to lateral changes in habitat (facies change in sedimentary strata), and that not all fossils may be found globally at the same time.

Study of the Composition of the Earth

Mineralogy

As a discipline, mineralogy has had close historical ties with geology. Minerals as basic constituents of rocks and ore deposits are obviously an integral aspect of geology. The problems and techniques of mineralogy, however, are distinct in many respects from those of the rest of geology, with the result that mineralogy has grown to be a large, complex discipline in itself.

Nepheline (greasy light gray), sodalite (blue), cancrinite (yellow), feldspar (white), and ferromagnesian minerals (black) in an alkalic syenite

About 3,000 distinct mineral species are recognized, but relatively few are important in the kinds of rocks that are abundant in the outer part of the Earth. Thus a few minerals such as the feldspars, quartz, and mica are the essential ingredients in granite and its near relatives. Limestones, which are widely distributed on all continents, consist largely of only two minerals, calcite and dolomite. Many rocks have a more complex mineralogy, and in some the mineral particles are so minute that they can be identified only through specialized techniques.

It is possible to identify an individual mineral in a specimen by examining and testing its physical properties. Determining the hardness of a mineral is the most practical way of identifying it. This can be done by using the Mohsscale of hardness, which lists 10 common minerals in their relative order of hardness: talc (softest with the scale number 1),gypsum (2), calcite (3), fluorite (4), apatite (5), orthoclase (6), quartz (7), topaz (8), corundum (9), and diamond (10). Harder minerals scratch softer ones, so that an unknown mineral can be readily positioned between minerals on the scale. Certain common objects that have been assigned hardness values roughly corresponding to those of the Mohs scale (e.g., fingernail, pocketknife blade, steel file) are usually used in conjunction with the minerals on the scale for additional reference.

Other physical properties of minerals that aid in identification are crystal form, cleavage type, fracture, streak, lustre, colour, specific gravity, and density. In addition, the refractive index of a mineral can be determined with precisely calibrated immersion oils. Some minerals have distinctive properties that help to identify them. For example, carbonate minerals effervesce with dilute acids; halite is soluble in water and has a salty taste; fluorite (and about 100 other minerals) fluoresces in ultraviolet light; and uranium-bearing minerals are radioactive.

The science of crystallography is concerned with the geometric properties and internal structure of crystals. Because minerals are generally crystalline, crystallography is an essential aspect of mineralogy. Investigators in the field may use a reflecting goniometer that measures angles between crystal faces to help determine the crystal system to which a mineral belongs. Another instrument that they frequently employ is the X-raydiffractometer, which makes use of the fact that X-rays, when passing through a mineral specimen, are diffracted at regular angles. The paths of the diffracted rays are recorded on photographic film, and the positions and intensities of the resulting diffraction lines on the film provide a particular pattern. Every mineral has its own unique diffraction pattern, so crystallographers are able to determine not only the crystal structure of a mineral but the type of mineral as well.

When a complex substance such as a magma crystallizes to form igneous rock, the grains of different constituent minerals grow together and mutually interfere, with the result that they do not retain their externally recognizable crystal form. To study the minerals in such a rock, the mineralogist uses a petrographic microscope constructed for viewing thin sections of the rock, which are ground uniformly to a thickness of about 0.03 millimetre, in light polarized by two polarizing prisms in the microscope. If the rock is crystalline, its essential minerals can be determined by their peculiar optical properties as revealed in transmitted light under magnification, provided that the individual crystal grains can be distinguished. Opaque minerals, such as those with a high content of metallic elements, require a technique employing reflected light from polished surfaces. This kind of microscopic analysis has particular application to metallic ore minerals. The polarizing microscope, however, has a lower limit to the size of grains that can be distinguished with the eye; even the best microscopes cannot resolve grains less than about 0.5 micrometre (0.0005 millimetre) in diameter. For higher magnifications the mineralogist uses an electron microscope, which produces images with diameters enlarged tens of thousands of times.

Another important area of mineralogy is concerned with the chemical composition of minerals. The primary instrument used is the electron microprobe. Here a beam of electrons is focused on a thin section of rock that has been highly polished and coated with carbon. The electron beam can be narrowed to a diameter of about one micrometre and thus can be focused on a single grain of a mineral, which can be observed with an ordinary optical microscope system. The electrons cause the atoms in the mineral under examination to emit diagnostic X-rays, the intensity and concentration of which are measured by a computer. Besides spot analysis, this method allows a mineral to be traversed for possible chemical zoning. Moreover, the concentration and relative distribution of elements such as magnesium and iron across the boundary of two coexisting minerals like garnet and pyroxene can be used with thermodynamic data to calculate the temperature and pressure at which minerals of this type crystallize.

Although the major concern of mineralogy is to describe and classify the geometrical, chemical, and physical properties of minerals, it is also concerned with their origin. Physical chemistry and thermodynamics are basic tools for understanding mineral origin. Some of the observational data of mineralogy are concerned with the behaviour of solutions in precipitating crystalline materials under controlled conditions in the laboratory. Certain minerals can be created synthetically under conditions in which temperature and concentration of solutions are carefully monitored. Other experimental methods include study of the transformation of solids at high temperatures and pressures to yield specific minerals or assemblages of minerals. Experimental data obtained in the laboratory, coupled with chemical and physical theory, enable the conditions of origin of many naturally occurring minerals to be inferred.

Petrology

Petrology is the study of rocks, and, because most rocks are composed of minerals, petrology is strongly dependent on mineralogy. In many respects mineralogy and petrology share the same problems; for example, the physical conditions that prevail (pressure, temperature, time, and presence or absence of water) when particular minerals or mineral assemblages are formed. Although petrology is in principle concerned with rocks throughout the crust, as well as with those of the inner depths of the Earth, in practice the discipline deals mainly with those that are accessible in the outer part of the Earth's crust. Rock specimens obtained from the surface of the Moon and from other planets are also proper considerations of petrology. Fields of specialization in petrology correspond to the aforementioned three major rock types—igneous, sedimentary, and metamorphic.

Igneous Petrology

Igneous petrology is concerned with the identification, classification, origin, evolution, and processes of formation and crystallization of the igneous rocks. Most of the rocks available for study come from the Earth's crust, but a few, such as eclogites, derive from the mantle. The scope of igneous petrology is very large because igneous rocks make up the bulk of the continental and oceanic crusts and of the mountain belts of the world, which range in age from early Archean to Neogene, and they also include the high-level volcanic extrusive rocks and the plutonic rocks that formed deep within the crust. Of utmost importance to igneous petrologic research is geochemistry, which is concerned with the major- and trace-element composition of igneous rocks as well as of the magmas from which they arose. Some of the major problems within the scope of igneous petrology are: (1) the form and structure of igneous bodies, whether they be lava flows or granitic intrusions, and their relations to surrounding rocks (these are problems studied in the field); (2) the crystallization history of the minerals that make up igneous rocks (this is determined with the petrographic polarizing microscope); (3) the classification of rocks based on textural features, grain size, and the abundance and composition of constituent minerals; (4) the fractionation of parent magmas by the process of magmatic differentiation, which may give rise to an evolutionary sequence of genetically related igneous products; (5) the mechanism of generation of magmas by partial melting of the lower continental crust, suboceanic and subcontinental mantle, and subducting slabs of oceanic lithosphere; (6) the history of formation and the composition of the present oceanic crust determined on the basis of data from the Integrated Ocean Drilling Program (IODP); (7) the evolution of igneous rocks through geologic time; (8) the composition of the mantle from studies of the rocks and mineral chemistry of eclogites brought to the surface in kimberlitepipes; (9) the conditions of pressure and temperature at which different magmas form and at which their igneous products crystallize (determined from high-pressure experimental petrology).

The basic instrument of igneous petrology is the petrographic polarizing microscope, but the majority of instruments used today have to do with determining rock and mineral chemistry. These include the X-ray fluorescence spectrometer, equipment for neutron activation analysis, induction-coupled plasma spectrometer, electron microprobe, ionprobe, and mass spectrometer. These instruments are highly computerized and automatic and produce analyses rapidly. Complex high-pressure experimental laboratories also provide vital data.

With a vast array of sophisticated instruments available, the igneous petrologist is able to answer many fundamental questions. Study of the ocean floor has been combined with investigation of ophiolite complexes, which are interpreted as slabs of ocean floor that have been thrust onto adjacent continental margins. An ophiolite provides a much deeper section through the ocean floor than is available from shallow drill cores and dredge samples from the extant ocean floor. These studies have shown that the topmost volcanic layer consists of tholeiitic basalt or mid-ocean ridge basalt that crystallized at an accreting rift or ridge in the middle of an ocean. A combination of mineral chemistry of the basalt minerals and experimental petrology of such phases allows investigators to calculate the depth and temperature of the magma chambers along the mid-ocean ridge. The depths are close to six kilometres, and the temperatures range from 1,150 °C to 1,279 °C. Comprehensive petrologic investigation of all the layers in an ophiolite makes it possible to determine the structure and evolution of the associated magma chamber.

In the island arcs and active continental margins that rim the Pacific Ocean, there are many different volcanic and plutonic rocks belonging to the calc-alkaline series. These include basalt; andesite; dacite; rhyolite; ignimbrite; diorite; granite; peridotite; gabbro; and tonalite, trondhjemite, and granodiorite (TTG). They occur typically in vast batholiths, which may reach several thousand kilometres in length and contain more than 1,000 separate granitic bodies. These TTG calc-alkaline rocks represent the principal means of growth of the continental crust throughout the whole of geologic time. Much research is devoted to them in an effort to determine the source regions of their parent magmas and the chemical evolution of the magmas. It is generally agreed that these magmas were largely derived by the melting of a subducted oceanic slab and the overlying hydrated mantle wedge. One of the major influences on the evolution of these rocks is the presence of water, which was derived originally from the dehydration of the subducted slab.

Sedimentary Petrology

The field of sedimentary petrology is concerned with the description and classification of sedimentary rocks, interpretation of the processes of transportation and deposition of the sedimentary materials forming the rocks, the environment that prevailed at the time the sediments were deposited, and the alteration (compaction, cementation, and chemical and mineralogical modification) of the sediments after deposition.

A banded-iron formation (BIF) rock recovered from the Temagami greenstone belt in Ontario, Canada, and dated to 2.7 billion years ago. Dark layers of iron oxide are intercalated with red chert.

There are two main branches of sedimentary petrology. One branch deals with carbonate rocks, na mely limestones and dolomites, composed principally of calcium carbonate (calcite) and calcium magnesium carbonate (dolomite). Much of the complexity in classifying carbonate rocks stems partly from the fact that many limestones and dolomites have been formed, directly or indirect ly, through the influence of organisms, including bacteria, lime-secreting algae, various shelled organisms (e.g., mollusks and brachiopods), and by corals. In limestones and dolomites that were deposited under marine conditions, commonly in shallow warm seas, much of the material initially forming the rock consists of skeletons of lime-secreting organisms. In many examples, this skeletal material is preserved as fossils. Some of the major problems of carbonate petrology concern the physical and biological conditions of the environments in which carbonate material has been deposited, including water depth, temperature, degree of illumination by sunlight, mo tion by waves and currents, and the salinity and other chemical aspects of the water in which deposition occurred.

The other principal branch of sedimentary petrology is concerned with the sediments and sedi mentary rocks that are essentially noncalcareous. These include sands and sandstones, clays and claystones, siltstones, conglomerates, glacial till, and varieties of sandstones, siltstones, and con glomerates (e.g., the graywacke-type sandstones and siltstones). These rocks are broadly known as clastic rocks because they consist of distinct particles or clasts. Clastic petrology is concerned with classification, particularly with respect to the mineral composition of fragments or particles, as well as the shapes of particles (angular versus rounded), and the degree of homogeneity of particle sizes. Other main concerns of clastic petrology are the mode of transportation of sedimen tary materials, including the transportation of clay, silt, and fine sand by wind; and the transpor tation of these and coarser materials through suspension in water, through traction by waves and currents in rivers, lakes, and seas, and sediment transport by ice.

Sedimentary petrology also is concerned with the small-scale structural features of sediments and sedimentary rocks. Features that can be conveniently seen in a specimen held in the hand are with in the domain of sedimentary petrology. These features include the geometrical attitude of mineral grains with respect to each other, small-scale cross stratification, the shapes and interconnections of pore spaces, and the presence of fractures and veinlets.

Instruments and methods used by sedimentary petrologists include the petrographic microscope for description and classification, X-ray mineralogy for defining fabrics and small-scale structures, physical model flume experiments for studying the effects of flow as an agent of transport and the development of sedimentary structures, and mass spectrometryfor calculating stable isotopes and the temperatures of deposition, cementation, and diagenesis. Wet-suit diving permits direct observation of current processes on coral reefs, and manned submersibles enable observation at depth on the ocean floor and in mid-oceanic ridges.

The plate-tectonic theory has given rise to much interest in the relationships between sedimen tation and tectonics, particularly in modern plate-tectonic environments—e.g., spreading-related settings (intracontinental rifts, early stages of intercontinental rifting such as the Red Sea, and late stages of intercontinental rifting such as the margins of the present Atlantic Ocean), mid-oceanic settings (ridges and transform faults), subduction-related settings (volcanic arcs, fore-arcs, back arcs, and trenches), and continental collision-related settings (the Alpine-Himalayan belt and late orogenic basins with molasse [i.e., thick association of clastic sedimentary rocks consisting chiefly of sandstones and shales]). Today many subdisciplines of sedimentary petrology are concerned with the detailed investigation of the various sedimentary processes that occur within these plate-tectonic environments.

Metamorphic Petrology

Metamorphism means change in form. In geology the term is used to refer to a solid-state recrystallization of earlier igneous, sedimentary, or metamorphic rocks. There are two main types of metamorphism: (1) contact metamorphism, in which changes induced largely by increase in temperature are localized at the contacts of igneous intrusions; and (2) regional metamorphism, in which increased pressure and temperature have caused recrystallization over extensive regions in mountain belts. Other types of metamorphism include local effects caused by deformation in fault zones, burning oil shales, and thrusted ophiolite complexes; extensive recrystallization caused by high heat flow in mid-ocean ridges; and shock metamorphism induced by high-pressure impacts of meteorites in craters on the Earth and Moon.

Metamorphic petrology is concerned with field relations and local tectonic environments; the description and classification of metamorphic rocks in terms of their texture and chemistry, which provides information on the nature of the premetamorphic material; the study of minerals and their chemistry (the mineral assemblages and their possible reactions), which yields data on the temperatures and pressures at which the rocks recrystallized; and the study of fabrics and the relations of mineral growth to deformation stages and major structures, which provides information about the tectonic conditions under which regional metamorphic rocks formed.

A supplement to metamorphism is metasomatism: the introduction and expulsion of fluids and elements through rocks during recrystallization. When new crust is formed and metamorphosed at a mid-oceanic ridge, seawater penetrates into the crust for a few kilometres and carries much sodium with it. During formation of a contact metamorphic aureole around a granitic intrusion, hydrothermal fluids carrying elements such as iron, boron, and fluorine pass from the granite into the wall rocks. When the continental crust is thickened, its lower part may suffer dehydration and form granulites. The expelled fluids, carrying such heat-producing elements as rubidium, uranium, and thorium migrate upward into the upper crust. Much petrologic research is concerned with determining the amount and composition of fluids that have passed through rocks during these metamorphic processes.

The basic instrument used by the metamorphic petrologist is the petrographic microscope, which allows detailed study and definition of mineral types, assemblages, and reactions. If a heating/freezing stage is attached to the microscope, the temperature of formation and composition of fluid inclusions within minerals can be calculated. These inclusions are remnants of the fluids that passed through the rocks during the final stages of their recrystallization. The electron microprobe is widely used for analyzing the composition of the component minerals. The petrologist can combine the mineral chemistry with data from experimental studies and thermodynamics to calculate the pressures and temperatures at which the rocks recrystallized. By obtaining information on the isotopic age of successive metamorphic events with a mass spectrometer, pressure–temperature–time curves can be worked out. These curves chart the movement of the rocks over time as they were brought to the surface from deep within the continental crust; this technique is important for understanding metamorphic processes. Some continental metamorphic rocks that contain diamonds and coesites (ultrahigh pressure minerals) have been carried down subduction zones to a depth of at least 100 kilometres (60 miles), brought up, and often exposed at the present surface within resistant eclogites of collisional orogenic belts—such as the Swiss Alps, the Himalayas, the Kokchetav metamorphic terrane in Kazakhstan, and the Variscan belt in Germany. These examples demonstrate that metamorphic petrology plays a key role in unraveling tectonic processes in mountain belts that have passed through the plate-tectonic cycle of events.

Economic Geology

The mineral commodities on which modern civilization is heavily dependent are obtained from the Earth's crust and have a prominent place in the study and practice of economic geology. In turn, economic geology consists of several principal branches that include the study of ore deposits, petroleum geology, and the geology of nonmetallic deposits (excluding petroleum), such as coal, stone, salt, gypsum, clay and sand, and other commercially valuable materials.

Coal barges on the Finow Canal.

The practice of economic geology is distinguished by the fact that its objectives are to aid in the exploration for and extraction of mineral resources. The objectives are therefore economic. In petroleum geology, for example, a common goal is to guide oil-well drilling programs so that the most profitable prospects are drilled and those that are likely to be of marginal economic value, or barren, are avoided. A similar philosophy influences the other branches of economic geology. In this sense, economic geology can be considered as an aspect of business that is devoted to economic decision making. Many deposits of economic interest, particularly those of metallic ores, are of extreme scientific interest in themselves, however, and they have warranted intensive study that has been somewhat apart from economic considerations.

The practice of economic geology provides employment for a large number of geologists. On a worldwide basis, probably more than two-thirds of those persons employed in the geologic sciences are engaged in work that touches on the economic aspects of geology. These include geologists whose main interests lie in diverse fields of the geologic sciences. For example, the petroleum industry, which collectively is the largest employer of economic geologists, attracts individuals with specialties in stratigraphy, sedimentary petrology, structural geology, paleontology, and geophysics.

Geochemistry

Chemistry of the Earth

Geochemistry is broadly concerned with the application of chemistry to virtually all aspects of geology. Inasmuch as the Earth is composed of the chemical elements, all geologic materials and most geologic processes can be regarded from a chemical point of view. Some of the major problems that broadly belong to geochemistry are as follows: the origin and abundance of the elements in the solar system, galaxy, and universe(cosmochemistry); the abundance of elements in the major divisions of the Earth, including the core, mantle, crust, hydrosphere, and atmosphere; the behaviour of ions in the structure of crystals; the chemical reactions in cooling magmas and the origin and evolution of deeply buried intrusive igneous rocks; the chemistry of volcanic (extrusive) igneous rocks and of phenomena closely related to volcanic activity, including hot-spring activity, emanation of volcanic gases, and origin of ore deposits formed by hot waters derived during the late stages of cooling of igneous magmas; chemical reactions involved in weathering of rocks in which earlier formed minerals decay and new minerals are created; the transportation of weathering products in solution by natural waters in the ground and in streams, lakes, and the sea; chemical changes that accompany compaction and cementation of unconsolidated sediments to form sedimentary rocks; and the progressive chemical and mineralogical changes that take place as rocks undergo metamorphism.

One of the leading general concerns of geochemistry is the continual recycling of the materials of the Earth. This process takes place in several ways: (1) It is widely believed that oceanic and continental basalts crystallized from magmas that were ultimately derived by partial melting of the Earth's mantle. Much geochemical research is devoted to the quantification of this extraction of mantle material and its contribution to crustal growth throughout geologic time in the many stages of seafloor formation and mountain building. (2) When the basalts that formed at the mid-oceanic ridge are transported across the ocean by the process of seafloor spreading, they interact with seawater, and this involves the adding of sodium to the basaltic crust and the extraction of calcium from it. (3) Geophysical data confirm the idea that the oceanic lithosphere is being consumed along the Earth's major subduction zones below the continental lithosphere—e.g., along the continental margin of the Andes Mountain Ranges. This may involve pelagic sediments from the ocean floor, oceanic basalts altered by seawater exchange, gabbros, ultramafic rocks, and segments of the underlying mantle. Many geochemists are studying what happens to this subducted material and how it contributes to the growth of island arcs and Andean-type mountain belts. (4) The behaviour of dissolved materials in natural waters, under the relatively low temperatures that prevail at or near the surface of the Earth, is an integral aspect of the crustal cycle. Weathering processes supply dissolved material, including silica, calcium carbonate, and other salts, to streams. These materials then enter the oceans, where some remain in solution (e.g., sodium chloride), whereas others are progressively removed to form certain sedimentary rocks, including limestone and dolomite, and, where conditions are conducive for the formation of deposits by means of evaporation, gypsum (hydrous calcium sulfate), rock salt (halite), and potash deposits may occur.

The behaviour of biological materials and their subsequent disposition are important aspects of geochemistry, generally termed organic geochemistry and biogeochemistry. Major problems of organic geochemistry include the question of the chemical environment on Earth in which life originated; the modification of the hydrosphere, and particularly the atmosphere, through the effects of life; and the incorporation of organic materials in rocks, including carbonaceous material in sedimentary rocks. The nature and chemical transformations of biological material present in deposits of coal, petroleum, and natural gaslie within the scope of organic geochemistry. Organic chemical reactions influence many geochemical processes, as, for example, rock weathering and production of soil, the solution, precipitation, and secretion of such dissolved materials as calcium carbonate, and the alteration of sediments to form sedimentary rocks. Biogeochemistry deals chiefly with the cyclic flows of individual elements and their compounds between living and nonliving systems.

Geochemistry has applications to other subdisciplines within geology, as well as to disciplines relatively far removed from it. At one extreme, geochemistry is linked with cosmology in a number of ways. These include the study of the chemical composition of meteorites, the relative abundance of elements in the Earth, Moon, and other planets, and the ages of meteorites and of rocks of the crust of the Earth and Moon as established by radiometric means. At the other extreme, the geochemistry of traces of metals in rocks and soils and, ultimately, in the food chain has important consequences for humans and for the vast body of lesser organisms on which they are dependent and with whom they coexist. Deficiencies in traces of copper and cobalt in forage plants, for example, lead to diseases in certain grazing animals and may locally influence human health. These deficiencies are in turn related to the concentrations of these elements in rocks and the manner in which they are chemically combined within soils and rocks.

The chemical analysis of minerals is undertaken with the electron microprobe . Instruments and techniques used for the chemical analysis of rocks are as follows: The X-ray fluorescent (XRF) spectrometer excites atoms with a primary X-ray beam and causes secondary (or fluorescent) X-rays to be emitted. Each element produces a diagnostic X-radiation, the intensity of which is measured. This intensity is proportional to the concentration of the element in the rock, and so the bulk composition can be calculated. The crushed powder of the rock is compressed into a disk or fused into a bead and loaded into the spectrometer, which analyzes it automatically under computer control. Analysis of most elements having concentrations of more than five parts per million is possible.

Neutron-activation analysis is based on the fact that certain elements are activated or become radiogenic when they are bombarded with a flux of neutrons formed from the radioactive decay of uranium-235 in a nuclear reactor. With the addition of the neutrons, the stable isotopes produce new unstable radionuclides, which then decay, emitting particles with diagnostic energies that can be separated and measured individually. The technique is particularly suitable for the analysis of the rare earth elements, uranium, thorium, barium, and hafnium, with a precision to less than one part per million.

The induction-coupled plasma (ICP) spectrometer can analyze over 40 elements. Here, a solution of a rock is put into a plasma, and the concentration of the elements is determined from the light emitted. This method is rapid, and the ICP spectrometer is particularly suited to analyzing large numbers of soil and stream sediment samples, as well as mineralized rocks in mineral exploration.

Isotopic Geochemistry

Isotopic geochemistry has several principal roles in geology. One is concerned with the enrichment or impoverishment of certain isotopic species that results from the influence of differences in mass of molecules containing different isotopes. Measurements of the proportions of various isotopic species can be used as a form of geologic thermometer. The ratio of oxygen-16 to oxygen-18 in calcium carbonate secreted by various marine organisms from calcium carbonate in solution in seawater is influenced by the temperature of the seawater. Precise measurement of the proportions of oxygen-16 with respect to oxygen-18 in calcareous shells of some fossilmarine organisms provides a means of estimating the temperatures of the seas in which they lived. The varying ocean temperatures during and between the major advances of glaciers during the ice ages have been inferred by analyzing the isotopic composition of the skeletons of floating organisms recovered as fossils in sediment on the seafloor. Other uses of isotopic analyses that involve temperature-dependent rate processes include the progressive removal of crystals from cooling igneous magmas.

Another role of isotopic geochemistry that is of great importance in geology is radiometric age dating. The ability to quantify the geologic timescale—i.e., to date the events of the geologic past in terms of numbers of years—is largely a result of coupling radiometric dating techniques with older, classical methods of establishing relative geologic ages. As explained earlier, radiometric dating methods are based on the general principle that a particular radioactive isotope (radioactive parent or source material) incorporated in geologic material decays at a uniform rate, producing a decay product, or daughter isotope. Some radiometric "clocks" are based on the ratio of the proportion of parent to daughter isotopes, others on the proportion of parent remaining, and still others on the proportion of daughter isotopes with respect to each other. For example, uranium-238 decays ultimately to lead-206, which is one of the four naturally occurring isotopic species of lead. Minerals that contain uranium-238 when initially formed may be dated by measuring the proportions of lead-206 and uranium-238; the older the specimen, the greater the proportion of lead-206 with respect to uranium-238. The decay of potassium-40 to form argon-40 (calcium-40 is produced in this decay process as well) is also a widely used radiometric dating tool, though there are several other parent-daughter pairs that are used in radiometric dating, including another isotope of uranium (uranium-235), which decays ultimately to form lead-207, and thorium-232, which decays to lead-208.

Uranium-238 and uranium-235 decay very slowly, although uranium-235 decays more rapidly than uranium-238. The rate of decay may be expressed in several ways. One way is by the radioactive isotope's half-life—the interval of time in which half of any given initial amount will have decayed. The half-life of uranium-238 is about 4,510,000,000 years, whereas the half-life of uranium-235 is about 713,000,000 years. Other radioactive isotopes decay at greatly differing rates, with half-lives ranging from a fraction of a second to quadrillions of years.

It is useful to combine a variety of isotopic methods to determine the complete history of a crustal rock. A samarium-147–neodymium-143 date on granitic gneiss, for example, may be interpreted as the time of mantle–crust differentiation or crustal accretion that produced the original magmatic granite. Also, a lead-207–lead-206 date on a zircon will indicate the crystallization age of the granite. In contrast, a rubidium-87–strontium-87 date of a whole rock sample may give the time at which the rock became a closed system for migration of the strontium during the period of metamorphism that converted the granite to granitic gneiss. When potassium-40 breaks down to argon-40, the argon continues to diffuse until the rock has cooled to about 200 °C; therefore, a potassium-40–argon-40 date may be interpreted as the time when the granite cooled through a blocking temperature that stopped all argon release. This may reflect the cooling of the granite during late uplift in a young mountain belt.

Since the 1980s two technological advancements have greatly increased the geologist's ability to compute the isotopic age of rocks and minerals. The SHRIMP (Sensitive High Mass Resolution Ion Microprobe) enables the accurate determination of the uranium-lead age of the mineral zircon, and this has revolutionized the understanding of the isotopic age of formationof zircon-bearing igneous granitic rocks. Another technological development is the ICP-MS (Inductively Coupled Plasma Mass Spectrometer), which is able to provide the isotopic age of zircon, titanite, rutile, and monazite. These minerals are common to many igneous and metamorphic rocks.

Carbon-14 is a radioactive isotope of carbon (carbon-12 and carbon-13 are stable isotopes) with a half-life of 5,570 years. Carbon-14 is incorporated in all living material, for it is derived either directly or indirectly from its presence in atmospheric carbon dioxide. The moderately short half-life of carbon-14 makes it useful for dating biological materials that are more than a few hundred years old and less than 30,000 years old. It has been used to provide correlation of events within this time span, particularly those of the Pleistocene Epoch involving the Earth's most recent ice ages.

Study of the Structure of the Earth

Geodesy

Earth scientists setting up equipment to monitor changes on the slopes
of Mount Saint Helens, Washington, U.S.

The scientific objective of geodesy is to determine the size and shape of the Earth. The practical role of geodesy is to provide a network of accurately surveyed points on the Earth's surface, the vertical elevations and geographic positions of which are precisely known and, in turn, may be incorporated in maps. When two geographic coordinates of a control point on the Earth's surface, its latitude and longitude, are known, as well as its elevation above sea level, the location of that point is known with an accuracy within the limits of error involved in the surveying processes. In mapping large areas, such as a whole state or country, the irregularities in the curvature of the Earth must be considered. A network of precisely surveyed control points provides a skeleton to which other surveys may be tied to provide progressively finer networks of more closely spaced points. The resulting networks of points have many uses, including anchor points or bench marks for surveys of highways and other civil features. A major use of control points is to provide reference points to which the contour lines and other features of topographic maps are tied. Most topographic maps are made using photogrammetric techniques and aerial photographs.

The Earth's figure is that of a surface called the geoid, which over the Earth is the average sea lev-el at each location; under the continents the geoid is an imaginary continuation of sea level. The geoid is not a uniform spheroid, however, because of the existence of irregularities in the attraction of gravity from place to place on the Earth's surface. These irregularities of the geoid would bring about serious errors in the surveyed location of control points if astronomical methods, which involve use of the local horizon, were used solely in determining locations. Because of these irregu-larities, the reference surface used in geodesy is that of a regular mathematical surface, an ellipsoid of revolution that fits the geoid as closely as possible. This reference ellipsoid is below the geoid in some places and above it in others. Over the oceans, mean sea level defines the geoid surface, but over the land areas the geoid is an imaginary sea-level surface.

Today perturbations in the motions of artificial satellites are used to define the global geoid and gravity pattern with a high degree of accuracy. Geodetic satellites are positioned at a height of 700–800 kilometres above the Earth. Simultaneous range observations from several laser sta-tions fix the position of a satellite, and radar altimeters measure directly its height over the oceans. Results show that the geoid is irregular; in places its surface is up to 100 metres higher than the ideal reference ellipsoid and elsewhere it is as much as 100 metres below it. The most like-ly explanation for this height variation is that the gravity (and density) anomalies are related to mantle convection and temperature differences at depth. An important observation that confirms this interpretation is that there is a close correlation between the gravity anomalies and the surface expression of the Earth's plate boundaries. This also strengthens the idea that the ultimate driving force of plate tectonics is a large-scale circulation of the mantle.

A similar satellite ranging technique is also used to determine the drift rates of continents. Re-peated measurements of laser light travel times between ground stations and satellites permit the relative movement of different control blocks to be calculated.

Geophysics

Geophysics pertains to studies of the Earth that involve the methods and principles of physics. The scope of geophysics touches on virtually all aspects of geology, ranging from considerations of the conditions in the Earth's deep interior, where temperatures of several thousands of degrees Celsius and pressures of millions of atmospheres prevail, to the Earth's exterior, including its at-mosphere and hydrosphere.

Professor Anne Hofmeister loading a rock sample into a laser-flash apparatus to measure the sample's thermal conductivity.

The study of the Earth's interior provides a good example of the geophysicist's approach to problems. Direct observation is obviously impossible. Extensive knowledge of the Earth's interior has been derived from a variety of measurements, however, including seismic waves produced by quakes that travel through the Earth, measurements of the flow of heat from the Earth's interior into the outer crust, and by astronomical and other geologic considerations.

Geophysics may be divided into a number of overlapping branches in the following way: (1) study of the variations in the Earth's gravity field; (2) seismology, the study of the Earth's crust and interior by analysis of the transmission of elastic waves that are reflected or refracted; (3) the physics of the outer parts of the atmosphere, with particular attention to the radiation bombardment from the Sun and from outer space, including the influence of the Earth's magnetic field on radiation intercepted by the planet; (4) terrestrial electricity, which is the study of the storage and flow of electricity in the atmosphere and the solid Earth; (5) geomagnetism, the study of the source, configuration, and changes in the Earth's magnetic field and the study and interpretation of the remanent magnetism in rocks induced by the Earth's magnetic field when the rocks were formed (paleomagnetism); (6) the study of the Earth's thermal properties, including the temperature distribution of the Earth's interior and the variation in the transmission of heat from the interior to the surface; and (7) the convergence of several of the above-cited branches for the study of the large-scale tectonic structures of the Earth, such as rifts, continental margins, subduction zones, mid-oceanic ridges, thrusts, and continental sutures.

The techniques of geophysics include measurement of the Earth's gravitational field using gravimeters on land and sea and artificial satellites in space; measurement of its magnetic field with hand-held magnetometers or larger units towed behind research ships and aircraft; and seismographic measurement of subsurface structures using reflected and refracted elastic waves generated either by earthquakes or by artificial means (e.g., underground nuclear explosions or ground vibrations produced with special pistons in large trucks). Other tools and techniques of geophysics are diverse. Some involve laboratory studies of rocks and other earth materials under high pressures and elevated temperatures. The transmission of elastic waves through the crust and interior of the Earth is strongly influenced by the behaviour of materials under the extreme conditions at depth; consequently, there is strong reason to attempt to simulate those conditions of elevated temperatures and pressures in the laboratory. At another extreme, data gathered by rockets and satellites yield much information about radiation flux in space and the magnetic effects of the Earth and other planetary bodies, as well as providing high precision in establishing locations in geodetic surveying, particularly over the oceans. Finally, it should be emphasized that the tools of geophysics are essentially mathematical and that most geophysical concepts are necessarily expounded mathematically.

Geophysics has major influence both as a field of pure science in which the objective is pursuit of knowledge for the sake of knowledge and as an applied science in which the objectives involve solution of problems of practical or commercial interest. Its principal commercial applications lie in the exploration for oil and natural gas and, to a lesser extent, in the search for metallic ore deposits. Geophysical methods also are used in certain geologic-engineering applications, as in determining the depth of alluvial fill that overlies bedrock, which is an important factor in the construction of highways and large buildings.

Much of the success of the plate tectonics theory has depended on the corroborative factual evidence provided by geophysical techniques. For example, seismology has demonstrated that the earthquake belts of the world demarcate the plate boundaries and that intermediate and deep seismic foci define the dip of subduction zones; the study of rockmagnetism has defined the magnetic anomaly patterns of the oceans; and paleomagnetism has charted the drift of continents through geologic time. Seismic reflection profiling has revolutionized scientific ideas about the deep structure of the continents: major thrusts, such as the Wind Riverthrust in Wyoming and the Moine thrust in northwestern Scotland, can be seen on the profiles to extend from the surface to the Moho at about 35-kilometres depth; the Appalachian Mountains in the eastern United States must have been pushed at least 260 kilometres westward to their present position on a major thrust plane that now lies at about 15 kilometres depth; the thick crust of Tibet can be shown to consist of a stack of major thrust units; the shape and structure of continental margins against such oceans as the Atlantic and the Pacific are beautifully illustrated on the profiles; and the detailed structure of entire sedimentary basins can be studied in the search for oil reservoirs.

Structural Geology

Structural geology deals with the geometric relationships of rocks and geologic features in general. The scope of structural geology is vast, ranging in size from submicroscopic lattice defects in crystals to mountainbelts and plate boundaries.

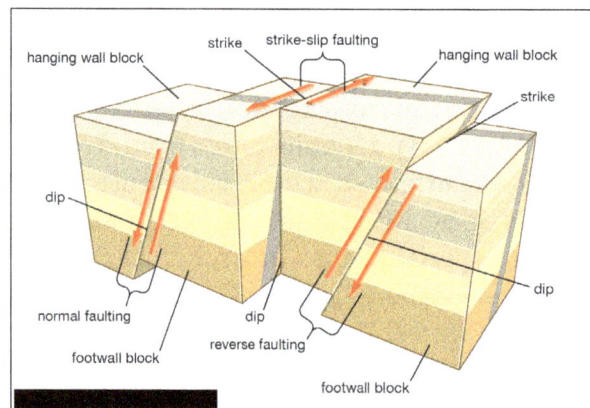

Types of faulting in tectonic earthquakes: In normal and reverse faulting, rock masses slip vertically past each other. In strike-slip faulting, the rocks slip past each other horizontally.

Structures may be divided into two broad classes: the primary structures that were acquired in the genesis of a rock mass and the secondary structures that result from later deformation of the primary structures. Most layered rocks (sedimentary rocks, some lava flows, and pyroclastic deposits) were deposited initially as nearly horizontal layers. Rocks that were initially horizontal may be deformed later by folding and may be displaced along fractures. If displacement has occurred and the rocks on the two sides of the fracture have moved in opposite directions from each other, the fracture is termed a fault; if displacement has not occurred, the fracture is called a joint. It is clear that faults and joints are secondary structures; i.e., their relative age is younger than the rocks that they intersect, but their age may be only slightly younger. Many joints in igneous rocks, for example, were produced by contraction when the rocks cooled. On the other hand, some fractures in rocks, including igneous rocks, are related to weathering processes and expansion associated with removal of overlying load. These will have been produced long after the rocks were formed. The faults and joints referred to above are brittle structures that form as discrete fractures within otherwise undeformed rocks in cool upper levels of the crust. In contrast, ductile structures result from permanent changes throughout a wide body of deformed rock at higher temperatures and pressures in deeper crustal levels. Such structures include folds and cleavage in slate belts, foliation in gneisses, and mineral lineation in metamorphic rocks.

The methods of structural geology are diverse. At the smallest scale, lattice defects and dislocations in crystals can be studied in images enlarged several thousand times with transmission electron microscopes. Many structures can be examined microscopically, using the same general techniques employed in petrology, in which sections of rock mounted on glass slides are ground very thin and are then examined by transmitted light with polarizing microscopes. Of course, some structures can be studied in hand specimens, which were preferably oriented when collected in the field.

On a large scale, the techniques of field geology are employed. These include the preparation of geologic maps that show the areal distribution of geologic units selected for representation on the map. They also include the plotting of the orientation of such structural features as faults, joints, cleavage, small folds, and the attitude of beds with respect to three-dimensional space. A common objective is to interpret the structure at some depth below the surface. It is possible to infer with some degree of accuracy the structure beneath the surface by using information available at the surface. If geologic information from drill holes or mine openings is available, however, the configuration of rocks in the subsurface commonly may be interpreted with much greater assurance as compared with interpretations involving projection to depth based largely on information obtained at the surface.

Strain analysis is another important technique of structural geology. Strain is change in shape; for example, by measuring the elliptical shape of deformed ooliths or concretions that must originally have been circular, it is possible to make a quantitative analysis of the strain patterns in deformed sediments. Other useful kinds of strain markers are deformed fossils, conglomerate pebbles, and vesicles. A long-term aim of such analysis is to determine the strain variations across entire segments of mountain belts. This information is expected to help geologists understand the mechanisms involved in the formation of such belts.

A combination of structural and geophysical methods are generally used to conduct field studies of the large-scale tectonic features. Field work enables the mapping of the structures at the surface, and geophysical methods involving the study of seismic activity, magnetism, and gravity make possible the determination of the subsurface structures.

The processes that affect geologic structures rarely can be observed directly. The nature of the deforming forces and the manner in which the Earth's materials deform under stress can be studied experimentally and theoretically, however, thus providing insight into the forces of nature. One form of laboratory experimentation involves the deformation of small, cylindrical specimens of rocks under very high pressures. Other experimental methods include the use of scale models of folds and faults consisting of soft, layered materials, in which the objective is to simulate the behaviour of real strata that have undergone deformation on a larger scale over much longer time.

Some experiments measure the main physical variables that control rock deformation—namely, temperature, pressure, deformation rate, and the presence of fluids such as water. These variables are responsible for changing the rheology of rocks from rigid and brittle at or near the Earth's surface to weak and ductile at great depths.

Tectonics

The subject of tectonics is concerned with the Earth's large-scale structural features. It forms a multidisciplinary framework for interrelating many other geologic disciplines, and thus it provides an integrated understanding of large-scale processes that have shaped the development of our planet. These structural features include mid-oceanic rifts; transform faults in the oceans; intracontinental rifts, as in the East African Rift Systemand on the Tibetan Highlands; wrench faults (e.g., the San Andreas Fault in California) that may extend hundreds of kilometres; sedimentary basins (oil potential); thrusts, such as the Main Central thrust in the Himalayas, that measure more than 2,000 kilometres long; ophiolite complexes; passive continental margins, as around the Atlantic Ocean; active continental margins, as around the Pacific Ocean; trench systems at the mouth of subduction zones; granitic batholiths (e.g., those in Sierra Nevada and Peru) that may be as long as 1,000 kilometres; sutures between collided continental blocks; and complete sections of mountain belts, such as the Andes, the Rockies, the Alps, the Himalayas, the Urals, and the Appalachians-Caledonians. Viewed as a whole, the study of these large-scale features encompasses the geology of plate tectonics and of mountain building at the margins of or within continents.

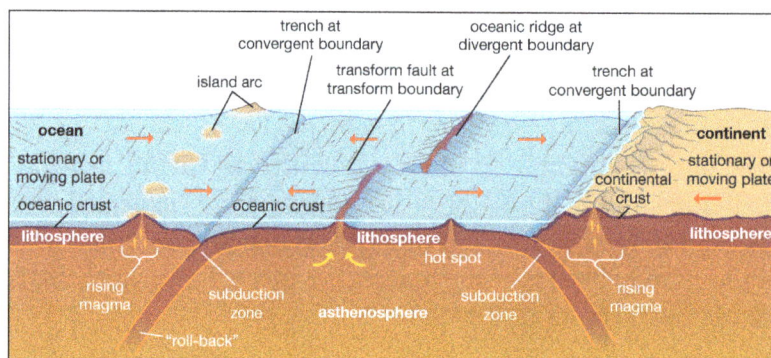

Crustal generation and destruction: Three-dimensional diagram showing crustal generation and destruction according to the theory of plate tectonics; included are the three kinds of plate boundaries—divergent, convergent (or collision), and strike-slip (or transform).

Volcanology

Volcanology is the science of volcanoes and deals with their structure, petrology, and origin. It is also concerned with the contribution of volcanoes to the development of the Earth's crust, with their role as contributors to the atmosphere and hydrosphere and to the balance of chemical elements in the Earth's crust, and with the relationships of volcanoes to certain forms of metallic ore deposits.

Many of the problems of volcanology are closely related to those of the origin of oceans and continents. Most of the volcanoes of the world are aligned along or close to the major plate boundaries, in particular the mid-oceanic ridges and active continental margins (e.g., the "Ring of Fire" around the Pacific Ocean). A few volcanoes occur within oceanic plates (e.g., along the Hawaiian chain);

these are interpreted as the tracks of plumes (ascending jets of partially molten mantle material) that formed when such a plate moved over hot spots fixed in the mantle.

Mayon Volcano: The 1984 eruption of Mayon Volcano, Luzon, Philippines.

One of the principal reasons for studying volcanoes and volcanic products is that the atmosphere and hydrosphere are believed to be largely derived from volcanic emanations, modified by biological processes. Much of the water present at the Earth's surface, which has aggregated mostly in the oceans but to a lesser extent in glaciers, streams, lakes, and groundwater, probably has emerged gradually from the Earth's interior by means of volcanoes, beginning very early in the Earth's history. The principal components of air—nitrogen and oxygen—probably have been derived through modification of ammonia and carbon dioxide emitted by volcanoes. Emissions of vapours and gases from volcanoes are an aspect of the degassing of the Earth's interior. Although the degassing processes that affect the Earth were probably much more vigorous when it was newly formed about 4,600,000,000 years ago, it is interesting to consider that the degassing processes are still at work. Their scale, however, is vastly reduced compared with their former intensity.

The study of volcanoes is dependent on a variety of techniques. The petrologic polarizing microscope is used for classifying lava types and for tracing their general mineralogical history. The X-ray fluorescence spectrometer provides a tool for making chemical analyses of rocks that are important for understanding the chemistry of a wide variety of volcanic products (e.g., ashes, pumice, scoriae, and bombs) and of the magmas that give rise to them. Some lavas are enriched or depleted in certain isotopic ratios that can be determined with a mass spectrometer. Analyses of gases from volcanoes and of hot springs in volcanic regions provide information about the late stages of volcanic activity. These late stages are characterized by the emission of volatile materials, including sulfurous gases. Many commercially valuable ore deposits have formed through the influence of hydrothermal volcanic solutions.

Volcanoes may pose a serious hazard to human life and property, as borne out by the destruction wrought by the eruptions of Mount Vesuvius (79 CE), Krakatoa, Mount Pelée , and Mount Saint Helens, to mention only a few. Because of this, much attention has been devoted to forecasting volcanic outbursts. In 1959 researchers monitored activity leading up to the eruption of Kilauea in Hawaii. Using seismographs, they detected swarms of earthquake tremors for several months prior to the eruption, noting a sharp increase in the number and intensity of small quakes shortly before the outpouring of lava. Tracking such tremors, which are generated by the upward movement of magma from the asthenosphere, has proved to be an effective means of determining the onset of eruptions and is now widely used for prediction purposes. Some volcanoes inflate when rising molten rock fills their magma chambers, and in such cases tiltmeters can be employed to detect a change in angle of the slope before eruption. Other methods of predicting violent volcanic activity involve the use of laser beams to check for changes in slope, temperature monitors, gas detectors, and instruments sensitive to variations in magnetic and gravity fields. Permanent volcano observatories have been established at some of the world's most active sites (e.g., Kilauea, Mount Etna, and Mount Saint Helens) to ensure early warning.

Study of Surface Features and Processes

Geomorphology

Geomorphology is literally the study of the form or shape of the Earth, but it deals principally with the topographical features of the Earth's surface. It is concerned with the classification, description, and origin of landforms. The configuration of the Earth's surface reflects to some degree virtually all of the processes that take place at or close to the surface as well as those that occur deep in the crust. The intricate details of the shape of a mountainrange, for example, result more or less directly from the processes of erosion that progressively remove material from the range. The spectrum of erosive processes includes weathering and soil-forming processes and transportation of materials by running water, wind action, and mass movement. Glacial processes have been particularly influential in many mountainous regions. These processes are destructional in the sense that they modify and gradually destroy the previous form of the range. Also important in governing the external shape of the range are the constructional processes that are responsible for uplift of the mass of rockfrom which the range has been sculptured. A volcanic cone, for example, may be created by the successive outpouring of lava, perhaps coupled with intermittent ejection of volcanic ash and tuff. If the cone has been built up rapidly, so that there has been relatively little time for erosive processes to modify its form, its shape is governed chiefly by the constructional processes involved in the outpouring of volcanic material. But the forces of erosion begin to modify the shape of a volcanic landform almost immediately and continue indefinitely. Thus, at no time can its shape be regarded as purely constructional or purely destructional, for its shape is necessarily a consequence of the interplay of these two major classes of processes.

Investigating the processes that influence landforms is an important aspect of geomorphology. These processes include the weathering caused by the action of solutions of atmospheric carbon dioxide and oxygen in water on exposed rocks; the activity of streams and lakes; the transport and deposition of dust and sand by wind; the movement of material through downhill creep of soil and rock and by landslides and mudflows; and shoreline processes that involve the mechanics and effects of waves and currents. Study of these different types of processes forms subdisciplines that exist more or less in their own right.

Glacial Geology

Glacial geology can be regarded as a branch of geomorphology, though it is such a large area of research that it stands as a distinct subdiscipline within the geologic sciences. Glacial geology is concerned with the properties of glaciers themselves as well as with the effects of glaciers as agents of both erosion and deposition. Glaciers are accumulations of snow transformed into solid ice. Important questions of glacial geology concern the climatic controls that influence the occurrence of glaciers, the processes by which snow is transformed into ice, and the mechanism of the flow of ice within glaciers. Other important questions involve the manner in which glaciers serve as erosive agents, not only in mountainous regions but also over large regions where great continental glaciers now extend or once existed. Much of the topography of the northern part of North America and Eurasia, for example, has been strongly influenced by glaciers. In places, bedrock has been scoured of most surficial debris. Elsewhere, deposits of glacial till mantle much of the area. Other extensive deposits include unconsolidated sediments deposited in former lakes that existed temporarily as a result of dams created by glacial ice or by glacial deposits. Many presently existing lakes are of glacial origin as, for example, the Great Lakes.

Research in glacial geology is conducted with a variety of tools. Investigators use, for example, radar techniques to determine the thickness of glaciers. In order to calculate the progressive advance or retreat of glacial masses, they ascertain the age of organic materials associated with glacial moraines by means of isotopic analyses.

Other branches of the geologic sciences are closely linked with glacial geology. In glaciated regions the problems of hydrology and hydrogeology are strongly influenced by the presence of glacial deposits. Furthermore, the suitability of glacial deposits as sites for buildings, roads, and other man-made features is influenced by the mechanical properties of the deposits and by soils formed on them.

Paleontology

The geologic time scale is based principally on the relative ages of sequences of sedimentary strata. Establishing the ages of strata within a region, as well as the ages of strata in other regions and on different continents, involves stratigraphic correlation from place to place. Although correlation of strata over modest distances often can be accomplished by tracing particular beds from place to place, correlation over long distances and over the oceans almost invariably involves comparison of fossils. With rare exceptions, fossils occur only in sedimentary strata. Paleontology, which is the science of ancient life and deals with fossils, is mutually interdependent with stratigraphy and with historical geology. Paleontology also may be considered to be a branch of biology.

Organic evolution is the essential principle involved in the use of fossils for stratigraphic correlation. It incorporates progressive irreversible changes in the succession of organisms through time. A small proportion of types of organisms has undergone little or no apparent change over long intervals of geologic time, but most organisms have progressively changed, and earlier forms have become extinct and, in turn, have been succeeded by more modern forms. Organisms preserved as fossils that lived over a relatively short span of geologic time and that were geographically widespread are particularly useful for stratigraphic correlation. These fossils are indexes of relative geologic age and may be termed index fossils.

Fossils play another major role in geology because they serve as indicators of ancient environments. Specialists called paleoecologists seek to determine the environmental conditions under which a fossil organism lived and the physical and biological constraints on those conditions. Did the organism live in the seas, lakes, or bogs? In what type of biological community did it live? What was its food chain? In short, what ecological niche did the organism occupy? Because oil and natural gas only accumulate in restricted environments, paleoecology can offer useful information for fossil fuel exploration.

Invertebrate Paleontology

One of the major branches of paleontology is invertebrate paleontology, which is principally concerned with fossil marine invertebrate animals large enough to be seen with little or no magnification. The number of invertebrate fossil forms is large and includes brachiopods, pelecypods, cephalopods, gastropods, corals and other coelenterates (e.g., jellyfish), bryozoans, sponges, various arthropods (invertebrates with limbs—e.g., insects), including trilobites, echinoderms, and many other forms, some of which have no living counterparts. The invertebrates that are used as index fossils generally possess hard parts, a characteristic that has fostered their preservation as fossils. The hard parts preserved include the calcareous or chitinous shells of the brachiopods, cephalopods, pelecypods, and gastropods, the jointed exoskeletons of such arthropods as trilobites, and the calcareous skeletons of frame-building corals and bryozoans. The vast variety of organisms lacking hard parts are poorly represented in the geologic record; however, they sometimes occur as impressions or carbonized films in finely laminated sediments.

Vertebrate Paleontology

Vertebrate paleontology is concerned with fossils of the vertebrates: fish, amphibians, reptiles, birds, and mammals. Although vertebratepaleontology has close ties with stratigraphy, vertebrate fossils usually have not been extensively used as index fossils for stratigraphic correlation, vertebrates generally being much larger than invertebrate fossils and consequently rarer. Fossil mammals, however, have been widely used as index fossils for correlating certain nonmarine strata deposited during the Paleogene Period (about 65.5 to 23 million years ago). Much interest in dinosaurs has arisen because of the evidence that they became extinct approximately 65.5 million years ago (at the Cretaceous-Tertiary, or Cretaceous-Paleogene, boundary) during the aftermath of a large meteorite or comet impact.

Micropaleontology

Micropaleontology involves the study of organisms so small that they can be observed only with the aid of a microscope. The size range of microscopic fossils, however, is immense. In most cases, the term micropaleontology connotes that aspect of paleontology devoted to the Ostracoda, a subclass of crustaceans that are generally less than one millimetre in length; Radiolaria, marine (typically planktonic) protozoans whose remains are common in deep ocean-floor sediments; and Foraminifera, marine protozoans that range in size from about 10 centimetres to a fraction of a millimetre.

Generally speaking, micropaleontology involves successive ranges of sizes of microscopic fossils down to organisms that must be magnified hundreds of times or more for viewing. The study of ultrasmall fossils is perhaps the fastest growing segment of contemporary paleontology and is dependent on modern laboratory instruments, including electron microscopes. It is an important aspect of oil and natural gas exploration. Microfossils, which are flushed up boreholes in the drilling mud, can be analyzed to determine the depositional environment of the underlying sedimentary rocks and their age. This information enables geologists to evaluate the reservoir potential of the rock (i.e., its capacity for holding gas or oil) and its depth. Ostracods and foraminifera occur in such abundance and in so many varieties and shapes that they provide the basis for a detailed classification and time division of Mesozoic and Cenozoic sediments in which oil may occur.

Filamentous and spheroidal microfossils are important in many Precambrian sediments such as chert. They occur in rocks as old as 3,500,000,000 years and are thus an important testimony of early life on Earth.

Paleobotany

Paleobotany is the study of fossil plants. The oldest widely occurring fossils are various forms of calcareous algae that apparently lived in shallow seas, although some may have lived in freshwater. Their variety is so profuse that their study forms an important branch of paleobotany. Other forms of fossil plants consist of land plants or of plants that lived in swamp forests, standing in water that was fresh or may have been brackish, such as the coal-forming swamps of the Late Carboniferous Period (from 320,000,000 to 286,000,000 years ago).

Palynology

Palynology deals with plant spores and pollen that are both ancient and modern and is a branch of paleobotany. It plays an important role in the investigation of ancient climates, particularly through studies of deposits formed during glacial and interglacial stages. Study of a sequence of spore- or pollen-bearing beds may reveal successive climatic changes, as indicated by changes in types of spores and pollen derived from different vegetative complexes. Spores and pollen are borne by the wind and spread over large areas. Furthermore, they tend to be resistant to decay and thus may be preserved in sediments under adverse conditions.

Astrogeology

Astrogeology is concerned with the geology of the solid bodies in the solar system, such as the asteroids and the planets and their moons. Research in this field helps scientists to better understand the evolution of the Earth in comparison with that of its neighbours in the solar system. This subject was once the domain of astronomers, but the advent of spacecraft has made it accessible to geologists, geophysicists, and geochemists. The success of this field of study has depended largely on the development of advanced instrumentation.

The U.S. Apollo program enabled humans to land on the Moon several times since 1969. Rocks were collected, geophysical experiments were set up on the lunar surface, and geophysical measurements were made from spacecraft. The Soyuz program of the Soviet Union also collected much geophysical data from orbiting spacecraft. The mineralogy, petrology, geochemistry, and geochronology of lunar rocks were studied in detail, and this research made it possible to work out the geochemical evolution of the Moon. The various manned and unmanned missions to the Moon resulted in many other accomplishments: for example, a lunar stratigraphy was constructed; geologic maps at a scale of 1:1,000,000 were prepared; the structure of the maria, rilles, and craters was studied; gravity profiles across the dense, lava-filled maria were produced; the distribution of heat-producing radioactive elements, such as uranium and thorium, was mapped with gamma-ray spectrometers; the Moon's internal structure was determined on the basis of seismographic records of moonquakes; the heat flow from the interior was measured; and the day and night temperatures at the surface were recorded.

Since the late 1960s, unmanned spacecraft have been sent to the neighbouring planets. Several of these probes were soft-landed on Marsand Venus. Soil scoops from the Martian surface have been chemically analyzed by an on-board X-ray fluorescence spectrometer. The radioactivity of the surface materials of both Mars and Venus have been studied with a gamma-ray detector, the isotopic composition of their atmospheres analyzed with a mass spectrometer, and their magnetic fields measured. Relief and geologic maps of Mars have been made from high-resolution photographs and topographical maps of Venus compiled from radar data transmitted by orbiting spacecraft. Photographs of Mars and Mercury show that their surfaces are studded with many meteorite craters similar to those on the Moon. Detailed studies have been made of the craters, volcanic landforms, lava flows, and rift valleys on Mars, and a simplified geologic-thermal history has been constructed for the planet.

By the mid-1980s the United States had sent interplanetary probes past Jupiter, Saturn, and Uranus. The craft transmitted data and high-resolution photographs of these outer planetary systems, including their rings and satellites.

This research has given increased impetus to the study of tektites, meteorites, and meteorite craters on Earth. The mineralogy, geochemistry, and isotopic age of meteorites and tektites have been studied in detail. Meteorites are very old and probably originated in the asteroid belt between Mars and Jupiter, while tektites are very young and most likely formed from material ejected from terrestrial meteorite craters. Many comparative studies have been made of the development and shapes of meteorite craters on Earth, the Moon, Mars, and Mercury. Space exploration has given birth to a new science—the geology of the solar system. The Earth can now be understood within the framework of planetary evolution.

Practical Applications

Exploration for Energy and Mineral Sources

Over the past century, industries have developed rapidly, populations have grown dramatically, and standards of living have improved, resulting in an ever-growing demand for energy and mineral resources. Geologists and geophysicists have led the exploration for fossil fuels (coal, oil, natural gas, etc.) and concentrations of geothermal energy, for which applications have grown in recent years. They also have played a major role in locating deposits of commercially valuable minerals.

Coal

The Industrial Revolution of the late 18th and 19th centuries was fueled by coal. Though it has been supplanted by oil and natural gas as the primary source of energy in most modern industrial nations, coal nonetheless remains an important fuel.

Coal-exploration geologists have found that coal was formed in two different tectonic settings: (1) swampy marine deltas on stable continental margins, and (2) swampy freshwater lakes in graben (long, narrow troughs between two parallel normal faults) on continental crust. Knowing this and the types of sedimentary rock formations that commonly include coal, geologists can quite readily locate coal-bearing areas. Their main concern, therefore, is the quality of the coal and the thickness of the coal bed or seam. Such information can be derived from samples obtained by drilling into the rock formation in which the coal occurs.

Oil and Natural Gas

During the last half of the 20th century, the consumption of petroleumproducts increased sharply. This led to a depletion of many existing oil fields, notably in the United States, and intensive efforts to find new deposits.

Crude oil and natural gas in commercial quantities are generally found in sedimentary rocks along rifted continental margins and in intracontinental basins. Such environments exhibit the particular combination of geologic conditions and rock types and structures conducive to the formation and accumulation of liquid and gaseous hydrocarbons. They contain suitable source rocks (organically rich sedimentary rocks such as black shale), reservoir rocks (those of high porosity and permeability capable of holding the oil and gas that migrate into them), and overlying impermeable rocks that prevent the further upward movement of the fluids. These so-called cap rocks form petroleum traps, which may be either structural or stratigraphic depending on whether they were produced by crustal deformation or original sedimentation patterns.

Petroleum geologists concentrate their search for oil deposits in such geologic settings, mapping both the surface and subsurface features of a promising area in great detail. Geologic surface maps show subcropping sedimentary rocks and features associated with structural traps such as ridges formed by anticlines during the early stages of folding and lineations produced by fault ruptures. Maps of this kind may be based on direct observation or may be constructed with photographs taken from aircraft and Earth-orbiting satellites, particularly of terrain in remote areas. Subsurface maps reveal possible hidden underground structures and lateral variations in sedimentary rock bodies that might form a petroleum trap. The presence of such features can be detected by various means, including gravity measurements, seismic methods, and the analysis of borehole samples from exploratory drilling.

Another method used by petroleum geologists in exploratory areas involves the sampling of surface waters from swamps, streams, or lakes. The water samples are analyzed for traces of hydrocarbons, the presence of which would indicate seepage from a subsurface petroleum trap. This geochemical technique, along with seismic profiling, is often used to search for offshore petroleum accumulations.

Once an oil deposit has actually been located and well drilling is under way, petroleum geologists can determine from core samples the depth and thickness of the reservoir rock as well as its porosity and permeability. Such information enables them to estimate the quantity of the oil present and the ease with which it can be recovered.

Although only about 15 percent of the world's oil has been exploited, petroleum geologists estimate that at the present rate of demand the supply of recoverable oil will last no more than 100 years. Because of this rapid depletion of conventional oil sources, economic geologists have explored oil shales and tar sands as potential supplementary petroleum resources. Extracting oil from these substances is, however, very expensive and energy-intensive. In addition, the extraction process (mining and chemical treatment) poses environmental challenges, especially in regions where it occurs. Even so, oil shales and tar sands are abundant, and advances in recovery technology may yet make them attractive alternativeenergy resources.

Geothermal Energy

Another alternate energy resource is the heat from the Earth's interior. The surface expression of this energy is manifested in volcanoes, fumaroles, steam geysers, hot springs, and boiling mud pools. Global heat-flow maps constructed from geophysical data show that the zones of highest heat flow occur along the active plate boundaries. There is, in effect, a close association between geothermal energy sources and volcanically active regions.

A variety of applications have been developed for geothermal energy. For example, public buildings, residential dwellings, and greenhouses in such areas as Reykjavík, Iceland, are heated with water pumped from hot springs and geothermal wells. Hot water from such sources also is used for heating soil to increase crop production (e.g., in Oregon) and for seasoning lumber (e.g., in parts of New Zealand). The most significant application of geothermal energy, however, is the generation of electricity.

Mineral Deposits

The distribution of commercially significant mineral deposits, the economic factors associated with their recovery, and the estimates of available reserves constitute the basic concerns of economic geologists. Because continued industrial development is heavily dependent on mineral resources, their work is crucial to modern society.

It has long been known that certain periods of Earth history were especially favourable for the concentration of specific types of minerals. Copper, zinc, nickel, and gold are important in Archean rocks; magnetite and hematite are concentrated in early Proterozoic banded-iron formations; and there are economic Proterozoic uranium reserves in conglomerates. These mineral deposits and a variety of others that developed throughout the Phanerozoic Eon can be related to specific types of plate-tectonic environments. Among the latter are copper, leads, and zinc in intracontinental rifts. An interesting discovery has been the remarkable concentrations of gold, iron, zinc, and copper in brine pools and sulfide-rich muds in the Red Sea and in the Salton Sea in southern California. In many countries copper, nickel, and chromium deposits occur in ophiolite complexes obducted onto the continents from the ocean floor; porphyry copper and molybdenum deposits are found in association with granodioritic intrusions; and tungsten and tin deposits occur in many granites. The correlation of these associations and distributions with periods of Earth history, on the one hand, and plate-tectonic settings, on the other, have enabled regional metallogenetic provinces to be defined, which have proved helpful in the search for ore deposits.

Earthquake Prediction and Control

No natural event is as destructive over so large an area in so short a time as an earthquake. Throughout the centuries earthquakes have been responsible not only for millions of deaths but also for tremendous damage to property and the natural landscape. If major earthquakes could be predicted, it would be possible to evacuate population centres and take other measures that could minimize the loss of life and perhaps reduce damage to property as well. For this reason earthquake prediction has become a major concern of seismologists in the United States, Russia, Japan, and China.

World seismicity patterns show that earthquakes tend to occur along active plate boundaries where there is subduction (Japan) or strike-slip motion (California) and along strike-slip faults. Investigators agree that much more has to be learned about the physical properties of rocks in fault zones before they are able to make use of changes in these properties to predict earthquakes, though the use of Global Positioning Systems (GPS) at satellite ground stations over the years is providing quantitative data on a millimetre scale concerning the relative movement of crustal blocks across seismic faults. Recent research has suggested that rocks may become strained shortly before an earthquake and affect such observable properties of the Earth's crust as seismic wave velocity and radon concentration. Leveling surveys and tiltmeter measurements have revealed that deformation in the fault zone just prior to an earthquake may cause changes in ground level and, in certain cases, variations in groundwaterlevel. Also, some investigators have reported changes in the electric resistivity and remanent magnetization of rocks as precursory phenomena.

Seismological research includes the study of earthquakes caused by human activities, such as impounding water behind high dams, injecting fluids into deep wells, excavating mines, and detonating underground nuclear explosions. In all of these cases except for deep mining, seismologists have found that the induction mechanism most likely involves the release of elastic strain, just as with earthquakes of tectonic origin. Studies of artificially induced quakes suggest that one possible method of controlling natural earthquakes is to inject fluids into fault zones so as to release strain energy.

Seismologists have done much to explain the characteristics of ground motions recorded in earthquakes. Such information is required to predict ground motions in future earthquakes, thereby enabling engineers to design earthquake-resistant structures. The largest percentage of the deaths and property damage that result from an earthquake is attributable to the collapse of buildings, bridges, and other man-made structures during the violent shaking of the ground. An effective way of reducing the destructiveness of earthquakes, therefore, is to build structures capable of withstanding intense ground motions.

Other Areas of Application

The fields of engineering, environmental, and urban geology are broadly concerned with applying the findings of geologic studies to construction engineering and to problems of land use. The location of a bridge, for example, involves geologic considerations in selecting sites for the supporting piers. The strength of geologic materials such as rock or compacted clay that occur at the sites of the piers should be adequate to support the load placed on them. Engineering geology is concerned with the engineering properties of geologic materials, including their strength, permeability, and compactability, and with the influence of these properties on the selection of locations for buildings, roads and railroads, bridges, dams, and other major civil features.

Urban geology involves the application of engineering geology and other fields of geology to environmental problems in urban areas. Environmental geology is generally concerned with those aspects of geology that touch on the human environment. Environmental and urban geology deal in large measure with those aspects of geology that directly influence land use. These include the stability of sites for buildings and other civil features, sources of water supply (hydrogeology), contamination of waters by sewage and chemical pollutants, selection of sites for burial of refuse so as to minimize pollution by seepage, and locating the source of geologic building materials, including sand, gravel, and crushed rock. Since the late 1990s the importance of environmental geology has increased considerably in most developed countries as societies became aware of the environmental impact of humankind.

Chapter 2

Geoscience: A Comprehensive Study

The branch of natural science which focuses on the physical constitution of the Earth and its atmosphere is called geoscience. There are a number of sub-disciplines within geoscience such as geobiology, geochemistry and geophysics. This chapter discusses in detail these sub-disciplines of geoscience.

Geoscience is the study of the Earth - its oceans, atmosphere, rivers and lakes, ice sheets and glaciers, soils, its complex surface, rocky interior, and metallic core. This includes many aspects of how living things, including humans, interact with the Earth. Geoscience has many tools and practices of its own but is intimately linked with the biological, chemical, and physical sciences.

The Disciplines of Geoscience

Geoscience is a complex and broad area of study. It is possible to study geoscience as an umbrella discipline (although it is usually called Earth Sciences at most colleges and universities across the world). Students can gain a broad view of many of the following areas. Understanding the sciences that come to make up the umbrella term can help us understand the interconnectedness of the geosciences. Few exist in a bubble; few have little to no impact on other areas within geoscience.

Archaeology and Anthropology

It may surprise some that the study of people in the past (anthropology) and human material remains from the past (archaeology) are geosciences. Increasingly, these disciplines are expanding outside of their traditional study areas. In the past, archaeologists dug up trinkets, curious artifacts and "treasures" (a subjective term considering that even if something is financially worthless, it can be culturally priceless). Anthropologists studied humans of the past to learn how those societies lived. Today, both are multidisciplinary, straddling social sciences and geosciences examining areas with contexts that would previously have seemed unrelated. One area it is now inextricably linked is in the environmental sciences.

Archaeologists and anthropologists today are interested in the study of the environmental past and its impact on such practices as agricultural development and resource gathering, geology for mineral mining such as flint for stone tools, and iron deposits for metallurgy. Their overlaps with geography for practices in the past such as topographical adaptation and how past societies used landscapes to express power, create a defensive position, or to best use it for weather patterns. There is even a subdiscipline called "geoarchaeology" which uses the tools and methods and applies the scientific principles of the geosciences to archaeology.

Archaeology also using tools used in a wide variety of other areas such as architecture - GIS and geophysics, cartography, aerial survey, satellite data and magnetic resonance imaging. Each of these tools is used in geological and geographical applications but are also vital to the modern archaeologist too.

Astronomy

Most people perceived astronomy as the study of planets, stars and celestial bodies. That is not untrue, but astronomy as a geoscience is largely concerned with how extraterrestrial bodies impact the planet's processes. Astronomers who look towards the Earth in their studies consider tidal ranges caused by the moon, Near Earth Objects and their impact on our planet when they pass by or are caught by our atmosphere. They will also seek to understand the geochemical or other changes that occur when an asteroid strikes the surface of the Earth. Astronomers overlap with many areas including atmospheric sciences (when astronomical events affect weather patterns - tidal ranges, the magnetosphere and even eclipses), palaeontology when looking at cataclysmic localized wide-scale extinction events and ecology.

Atmospheric Sciences

From one of the least obvious to perhaps one of the most obvious disciplines that come under geoscience. Atmospheric science covers anything to do with the atmosphere from localized and short-term weather patterns (meteorology) to seasonal climatic events such as El Nino, La Nina, Trade Winds, and long-term climatic changes caused by aerial pollutants or natural processes (climatology/climate science). Experts in this area will also need a broad understanding of issues such as chemistry, physics and the dynamics of all planetary processes that impact the atmosphere short and long-term. However, they also require an understanding of extraterrestrial phenomena such as the processes of the sun and how it impacts atmospheric processes and the impact of aerial bodies such as asteroids, meteorites, comets and bolides.

The atmosphere is fundamental to many other geosciences. Weather patterns change with seasons, but we also have unseasonal weather and these require explanation and study. Atmospheric scientists examine these phenomena to look for trends or to understand why they are happening the way they are. This area also covers meteorology (the study of the weather) and makes predictions for agriculture, the military, government and other areas. They are at the forefront of disaster relief and planning, using many techniques and technologies, some of which overlap with other disciplines. For example, satellite imagery and digital mapping.

Ecology

Ecology is the study of biological systems. It surprises many people to learn that the interaction of gut flora in the human digestive system is also ecology. However, in the context of geoscience, ecology is the study of the interactions of organisms within a localized, regional or global environment. A wide-open area of tundra in Alaska is ecology, but so is 2-3 acres of heath in a small town or Central Park as an island of green in an urban environment. Arguably, so is a town or city regardless of its size. The balance between the biological systems that share it and the study of it and them is "ecology". Those who study and work in ecology as a geoscience will examine a wide range of issues from biodiversity, species distribution and numbers, relationships between competing groups and between predators and prey, as well as the human impact on these aspects.

Some may also be interested in the nutrient cycle and ensuring it remains in balance, food supply, the impact of weather and climate on ecology, life cycles of organisms, and sometimes genetics. It is not mistakenly assumed it concerns environmentalism, but it is not, nor is it concerned with natural history. Its actual overlap is with genetics and evolutionary biology, with zoology and animal behaviors although its findings will have practical applications to these areas.

Environmental Science

Environmental science is the study of environments, the biological entities that inhabit it, the processes that affect it and human interaction. It is about studying all the aspects of the world in examining how the world works. In some ways, it overlaps with ecology and atmospheric sciences, (particularly in areas such as climate science) but it differs in many ways too. Firstly, environmental science is an applied science that uses the evidence from most of the disciplines listed here in the application of finding solutions to environmental problems or in understanding environmental processes. Experts in this area use a broad multidisciplinary approach to an interdisciplinary field. Yet it is also an explanation and a tool for these other sciences. It's the study of the effects of natural processes on the environment, including ecology, atmospheric sciences, oceanography and seismology. This incorporates not just the biological (as in ecology) but also the physical, chemical, mineralogical, atmospheric, and geographical (including geology) in its application.

Geography

Geography is the area of geoscience concerned with describing and explaining the profile of the planet we inhabit. Geographers examine land masses, features on land and at sea, the people who inhabit these spaces, and the various planetary atmospheric and geological phenomena. It is broadly broken down into two areas:

- Physical Geography: the study of natural physical processes of the planet from geosphere (sub-surface) through the hydrosphere (water) biosphere (life) to the upper atmosphere).

- Human Geography: the examination of people and communities, cultural practices within a landscape, and landscape economics.

The study of modern geography is not limited to our own planet. Many works with astronomers and astrophysicists in studying what we have learnt in the Space Age about our nearest planetary and lunar neighbors.

Geography is truly a universal scientific discipline examining both the hard sciences and the social sciences, in some ways marrying them together. As a broad area, it has four traditions under its umbrella which include spatial analyses - both natural and anthropogenic, the study of place and regionality, relationships between people and the land, and of course - geography as an Earth science/geoscience. It can inform and crossover with other disciplines. Human geography lends itself to landscape archaeology, for example. Geography also concerns the findings and processes of geology, limnology, seismology and volcanology, and ecology. Without geography, there would be no cartography, GIS, surveying, satellite information, aerial survey, geomatics or many other tools of the sciences.

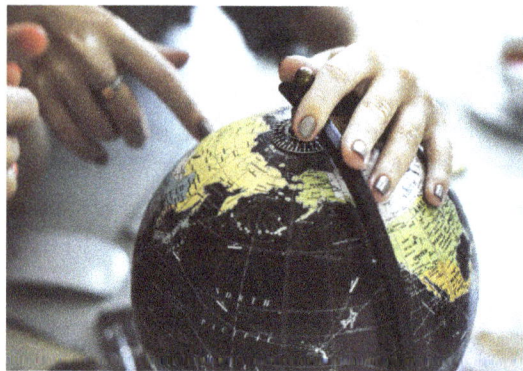

Geology

To many, geology is the study of rocks but that only covers one area of the academic discipline. Geology is actually the study of planetary physical structures and associated processes. This does include rock formation and stratigraphy, but it also concerns plate tectonics which examines mountain formation, and earthquakes and volcanoes, but also examines how the planet's structure has changed over time. They are also interested in the processes that create short-term or localized weather phenomena such as flooding, landslide and avalanches as well as erratic weather and the effects of geological events on the weather. For example, they might use their understanding of mountain formation to explain the impact of mountain ranges on the weather, and microclimates in valleys or on high land.

Geology overlaps with other disciplines of geoscience such as palaeontology and paleobotany, and even archaeology when considering natural processes of the Earth. These seemingly disparate areas of study share tools and methods as well as an interest in the study of the past profile of the planet. Sometimes, geological events lead to localized climatic changes that can sometimes cause local ecological changes or even widespread extinction. In 2018, a seminal academic conference brought together world experts in geology to examine the key evidence and indicators for the Great Extinctions.

Geomorphology

Another discipline related to geology, geomorphology is the examination of land - the form and shape, how it came to take that particular form or shape, the processes that alter it and the examination of deposits and erosion of geological material over time. The land is shaped by meteorological, hydrological and geological processes - air (winds), water (as solid ice or liquid) shapes a landscape through short-term or long-term climatic events. Geomorphologists examine a wide range of topographies from ocean trenches to deserts, to hill and mountains, the processes of glaciation and melt. They also use cartography and GIS to map, trace and monitor the spread of these features over geological time.

They also make no distinction between long-term climate change (desertification for example) as agents of geological morphological change and single events (flooding or mudslide that fundamentally changes the course of a river by destroying cliffs), nor between localized events and larger geological issues that can cause widespread problems such as volcanic explosions. Some refer to them as "landscape detectives" as they see landform as puzzles for solving asking questions such as "why is the landscape like this?" and "what processes led to it taking this particular ecological or geological form?" Recently, such questions have been applied to the surface of Mars.

Glaciology

A niche area but one vital for the geosciences especially as we come to understand Ice Age Theory, Glaciology is the scientific examination of ice as a natural phenomenon, typically with a focus on glaciation, but also looking at seasonal changes in ice and frost levels in areas of the world that expect to get seasonal ice. This area is vital because the retreat of Arctic and Antarctic glaciers is already impacting global climate. Through this area of study, we know of the potential impacts of melting ice, for example, sea level rises, but also the havoc the massive increase in cooler water (and diluted salts) will play on the ocean jet streams that keep areas of the North Pacific and North Atlantic temperate.

This is an interdisciplinary and multidisciplinary area, integrating geology and geography, geomorphology and hydrology, geophysics and geodesy, and biology and ecology. However, it also has uses for archaeologists and anthropologists studying humans during the last Ice Age, the spread of stone tools and adaptation to increasing and retreating ice. It's useful for palaeontology in the study of climate change and the extinction of species such as mammoth, woolly rhino and even human ancestor species such as Neanderthals that went extinct at the end of the last Ice Age. Understanding the science of Ice Ages through paleoclimate data could also help predict and mitigate modern era extinction.

Hydrogeology

Hydrogeology is an area of study grouped with geology that examines the physical structures of our planet how it relates to water processes - as a solid, liquid and gas. This means the physical actions of water, its present and past motion, and geological indicators of past water presence - therefore it's as much about water behavior as it is about water's physical properties relating to geology. This is a complex science as lakes, rivers, seas and oceans do not always follow dry land topography. Other geological processes also affect how water acts and reacts - for example, volcanic and seismic activities such as geothermal processes that create geysers, springs (cold and hot).

The study of past water presence is particularly useful to understanding geological history. It was such a process of understanding water flow that led to the conclusion that there was once free-flowing water on the surface of Mars and also the theory that water deposits exist beneath the surface of that planet locked up in extensive cave systems. It ties in closely with glaciology when concerned with examining how glaciers shape and mold landscapes, for example in the creation of mountains and valleys - most notable in North America and northern Europe.

Hydrology

Hydrology is the study of water - its physical properties, movement, quality, distribution, chemical makeup and its states (gas, liquid, solid) at any place on our planet from the deep oceans to the very limits of the upper atmosphere. It is less concerned with its impact on the geological structure of our planet. We know how important water is for life; the first complexes life forms were aquatic and only later did their descendants evolve to be able to cope with the chemical makeup of the atmosphere on dry land. The amount of water available to us is as much a scientific problem as an environmental engineering problem. We need it to drink, to water our crops and livestock for food; it also has commercial and industrial uses.

Much of hydrology today is about ensuring water and food security, sustainability in a growing population, and conserving it for planetary ecology. Called "water budget" it attempts to put a figure on how much is required and how much we can use as a community, as a country, or as a planet. Hydrologists may also study water pollution in our rivers, lakes and oceans and raise awareness of such problems.

Limnology

On a similar theme to hydrology and hydrogeology, limnology is the study of water bodies as inland features. This means streams, rivers, ponds, lakes, springs, and wetlands, and landlocked seas, either saline or freshwater but not oceans (that would be oceanography). The difference is that the bodies that come under limnology have a direct impact on and are impacted by the dry land ecologies in which they are located; open oceans do not share such a relationship. This area of research concerns the biological, physical, chemical and geological aspects of such water bodies. It's closely related to a number of other sciences that come under the geoscience banner but many now consider it a subdiscipline of ecology.

All aspects of these water bodies are of interest to limnologists including how the water flows, its chemical and mineral makeup, food and nutrient cycle (including bacteria, viruses and algae), pollution issues, ecosystems, drainage, relationships with land and other water bodies, and human interaction. It is both a theoretical and an applied science.

Mineralogy

Another subdiscipline of geology but standing on its own in many important ways, mineralogy is the study of the crystal structures, chemical properties, optical attributes and many other aspects of minerals. Earth scientists who specialize in mineralogy might look at how such minerals form, their geographical location, topographical indicators and their potential industrial and commercial applications. Typically, they will examine the presence of metals (precious or otherwise), precious stones, but also naturally occurring substances such as clay and gypsum. Each of these materials has a monetary value or a practical use. For example, clay varies geographically with some types of clay suited to certain types of firing. Not all clay types are useful in household crockery due to the temperature firing range. Gypsum is a useful building material and its quality determines its potential use when mining.

Oceanography

This is the area of Earth Sciences or Geoscience concerned with the oceans and everything in them. This covers ocean geology (including plate tectonics), oceanic ecosystems, chemical composition, wildlife and natural processes such as the water budget and jet streams. They also look at how the oceans function globally, weather systems and patterns and their effects on every other aspect of oceanography for water bodies that are not landlocked, but connected to other seas and oceans. It is the opposite of limnology which is the study of landlocked water bodies (both saline and fresh-water). Most oceanographers will specialize in one area when they finish a course of study, entering fields such as ocean geology or marine biology.

This area may also be subdivided into chemical oceanographers (who study chemical balances and its effect on the ecosystem such as coral bleaching); biological oceanography who study the wildlife of the various oceans; physical oceanographers examine the physics of the natural processes of the oceans such as wave systems, jet streams and so on. Then there are the oceanography geologists who examine plate tectonics and the physical structure of the planet beneath the oceans.

Paleontology

Like archaeology many are surprised to see that paleontology and its various subdisciplines (paleobotany, invertebrate paleontology, and micropaleontology) is now classified as an Earth science/geoscience. It once was about digging up curious fossils of extinct species, But today, due to its extracurricular interests in ecology, geography, geology, climate science (and paleoclimate studies), topography and other aspects of the natural world, it's firmly one where its adherents take into account the natural world in context.

Just as archaeologists are interested in past environments and environmental change, so are paleontologists. They also use similar tools to archaeologists and anthropologists such as satellite data and spatial studies to determine geographical distribution, applied using GIS and geophysics. Many of our top colleges and universities combine resources, methods and tools when studying paleontology in a geological context. Most species often span geological time and are accessed through research and study of sedimentary rock. Geological changes (such as plate tectonic shifts) are equally important in piecing together how the continents may have fitted together hundreds of millions of years ago. The distribution of paleontological remains across ocean boundaries is another piece in the puzzle of how the world may have looked long before the arrival of the first primates from which humans would eventually evolve.

Petrology

Closely related to geology within the range of geosciences, this is the area that examines how and why certain types of rock form in the way that they do. Rock types are broken down into igneous (of volcanic rock), metamorphic (when a chemical process changes one rock type to another), and sedimentary (originating from the build-up and compression of organic material such as silt, dead plants and microfossils) rocks. This area of study is particularly useful in fossil fuel discovery and use, as well as an examination of the properties of the different type of rock. It utilizes other areas of geology such as mineralogy and geochemical studies to explain rock texture, composition and properties, or why they ended up shaped a certain way. A petrologist may be interested in the rock layers of the Grand Canyon and explain the stratigraphic formation.

But now there is a fourth branch. Experimental petrology seeks to discover the potential uses of rock types and their applications by testing theories based on their physical and chemical properties. This area may lead to the production of renewable (and possibly cleaner with the right investment) synthetic fossil fuels. However, it also has academic uses in the study of rock formations in the upper and lower mantle of our planet, an area difficult to study rock properties in situ.

Seismology

This is the geoscience concerned with planetary elastic vibration, also known as "earthquakes". Seismologists study the appearance of earthquakes, predict them, and study their form and results as well as looking at the structures that can cause them. It is as much about the ripple effects and the aftereffects, including aftershock profile as it is about prediction (which is the public face of this area of the science). Seismologists use a variety of electronic survey equipment as well as applied scientific methods, particularly in math and physics.

Earthquakes are an ongoing problem and a matter of curiosity in examining the small geological events that can shape the environment. Largely concerned with plate tectonics, lately it's included wider environmental effects including destruction of natural habitats such as forest, localized or regional flooding through tsunamis, and the wider impact on the local and global atmosphere.

It's a relatively young science although humanity's interest in earthquakes goes back several millennia. The first instruments capable of predicting earthquakes were invented in the 1860s. Today, the science can be broken down into further subfields. One of which is paleoseismology which is the study of ancient earthquakes, how they have shaped environments, and allows researchers to make better predictions and more thorough explanations of their long-term effects.

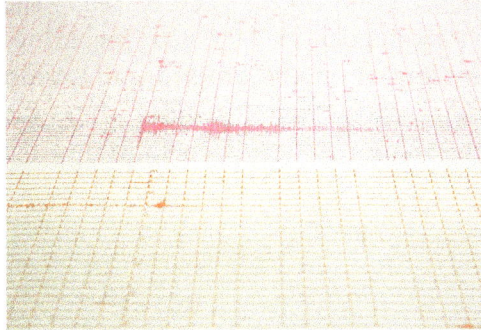

Soil Science

Soil science studies the Earth's surface natural resources. It covers everything from the nutrient cycle and soil formation, its regional physical and chemical properties, its relationship to fauna and flora, but also classification and soil engineering. Some are also interested in soil regionality and cartography, sometimes for agricultural purposes (some crops won't grow in certain soil profiles) such as acid/alkali, fertility, depth and quality. Soil science is broken down into two areas: pedology and edaphology. The first is the study of the form and formation, chemistry and classification of soils. The latter is the examination of soil ecology that is its relationships with plants, trees and shrubs.

Soil science is becoming increasingly important in the developed and developing world. We need to grow enough food for our increasing population and to do so sustainably. Preserving soils in so-called "marginal landscapes" (land of poor agricultural quality) is just one issue - use of pesticides and herbicides and their impact on soils, sustainable water use, and pollution and soil fertility have raised concerns about how to preserve soil are ongoing too. Also, the results of flooding and drought and general food consumption are current problems soil science is presently engaged in tackling.

Speleology

Speleology is the academic study of caves and cave systems. Particularly, researchers examine how they form and why. They also examine the types of rock in which they form and the various processes that lead to it. It was once a simple curiosity of geology, but it has become so separated and unique that it's considered a whole area of study in itself. The modern speleologist needs to understand geology, chemistry and physics, but also understand the importance of cartography (for mapping cave systems). There is also overlap with biology as some species adapt to living in cave systems, species such as blind cave fish, and archaeology and anthropology as caves represent some of human ancestry's earliest "homes". Discovery and conservation of cave art is as much interest to speleology as it is to archaeology. Also, caves have been the source of some of our best sources of fossils - therefore, it is also useful to paleontology also.

Volcanology

Volcanology is often paired with seismology in academic discussions as the two share similarities (for example, both concern shifting plate tectonics and geological weaknesses that lead to each subsequent catastrophic geological phenomena). However, the two disciplines differ in many ways. Volcanology studies something that been physically seen - the creation of volcanoes and its ejecta including magma, lava, rock deposit and the ash (including its environmental impact). A volcanologist examines both the formation and eruptions of existing volcanoes. They might also be interested in historic (including prehistoric) episodes of eruptions. Volcanology uses similar equipment in monitoring ground tremors to predict a volcanic eruption, but that is largely where they part.

Geobiology

Geobiology is a scientific discipline in which the principles and tools of biology are applied to studies of the Earth. In concept, geobiology parallels geophysics and geochemistry, two longer established disciplines within the Earth sciences.

Geobiology is predicated on the observation that biological processes interact with physical processes at and near the Earth's surface. Take, for example, carbon, the defining element of life. Within the biosphere – the sum of all environments that support life on Earth – carbon exists in a number of forms and in several key reservoirs. It is present as CO_2 in the atmosphere; as CO_2, HCO_3^- and CO_3^{2-} dissolved in fresh and marine waters; as carbonate minerals in soils, sediments and rocks; and as a huge variety of organic molecules in organisms, in sediments and soils, and dissolved in lakes and oceans. Physical processes move carbon from one reservoir to another; for example, volcanoes add CO_2 to the atmosphere and chemical weathering removes it. Biological processes do as well. In two notable examples, photosynthesis reduces CO_2 to sugar, and respiration oxidizes organic molecules to CO_2. Since the industrial revolution, humans have oxidized sedimentary organic matter (by burning fossil fuels) at rates much higher than those characteristic of earlier epochs, making us important participants in the Earth's carbon cycle.

Other biologically important elements also cycle through the biosphere. Sulfur, nitrogen, and iron all link the physical and biological Earth, interacting with each other and, importantly, with the carbon cycle. And oxygen, key to environments that support large animals, including humans, is regulated by a complex and incompletely understood set of processes that, again, have both biological and physical components.

Unlike physical processes, life evolves, and so the array of biological processes in play within the biosphere has changed through time. The state of the environment supporting biological communities has changed as well. Indeed, given the close relationship between environment and population distributions on the present day Earth, it is reasonable to hypothesize that evolving life has significantly influenced the chemical environment through time and, conversely, that environmental change has influenced the course of evolution.

While metabolism encompasses many of the biological cogs in the biosphere, other processes also play important roles. For example, many organisms precipitate minerals, either indirectly by altering local chemical environments, or directly by building mineralized skeletons. Today, skeletons dominate the deposition of carbonate and silica on the seafloor, although this was not true before the evolution of shells, spicules and tests. More subtly, organisms interact with clays and other minerals in a series of surface interactions that are only now beginning to be understood. While much of geobiology focuses on chemical processes, organisms influence the Earth through physical activities as well – think of microbial communities that can stabilize sand beds or worms that irrigate sediments as they burrow. The example of burrowing reminds us that while microorganisms garner much geobiological attention, plants and animals also act as geobiological agents, and have done so for more than 500 million years.

In short, Earth surface processes once considered to be largely physical in nature – for example weathering and erosion – are now known to have key biological components. Life plays a critical role in the Earth system.

The Biological Carbon Cycle

All living things are made of elements, the most abundant of which are, oxygen, carbon, hydrogen, nitrogen, calcium, and phosphorous. Of these, carbon is the best at joining with other elements to form compounds necessary for life, such as sugars, starches, fats, and proteins. Together, all these forms of carbon account for approximately half of the total dry mass of living things.

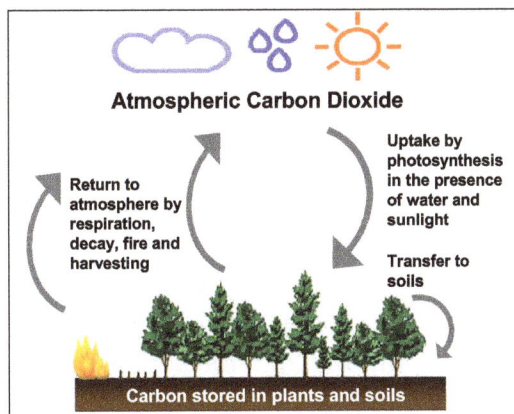

Atmospheric Carbon Dioxide

Uptake by photosynthesis in the presence of water and sunlight

Return to atmosphere by respiration, decay, fire and harvesting

Transfer to soils

Carbon stored in plants and soils

A sub-cycle within the global carbon cycle. Carbon continuously moves between the atmosphere, plants and soils through photosynthesis, plant respiration, harvesting, fire and decomposition.

Carbon is also present in the Earth's atmosphere, soils, oceans, and crust. When viewing the Earth as a system, these components can be referred to as carbon pools (sometimes also called stocks or reservoirs) because they act as storage houses for large amounts of carbon. Any movement of carbon between these reservoirs is called a flux. In any integrated system, fluxes connect reservoirs together to create cycles and feedbacks. An example of such a cycle is seen in Figure where, carbon in the atmosphere is used in photosynthesis to create new plant material. On a global basis, this processes transfers large amounts of carbon from one pool (the atmosphere) to another (plants). Over time, these plants die and decay, are harvested by humans, or are burned either for energy or in wildfires. All of these processes are fluxes that can cycle carbon among various pools within ecosystems and eventually releases it back to the atmosphere. Viewing the Earth as a whole, individual cycles like this are linked to others involving oceans, rocks, etc. on a range of spatial and temporal scales to form an integrated global carbon cycle.

On the shortest time scales, of seconds to minutes, plants take carbon out of the atmosphere through photosynthesis and release it back into the atmosphere via respiration. On longer time scales, carbon from dead plant material can be incorporated into soils, where it might reside for years, decades or centuries before being broken down by soil microbes and released back to the atmosphere. On still longer time scales, organic matter1 that became buried in deep sediments (and protected from decay) was slowly transformed into deposits of coal, oil and natural gas, the fossil fuels we use today. When we burn these substances, carbon that has been stored for millions of years is released once again to the atmosphere in the form of carbon dioxide (CO_2).

The carbon cycle has a large effect on the function and wellbeing of our planet. Globally, the carbon cycle plays a key role in regulating the Earth's climate by controlling the concentration of carbon dioxide in the atmosphere. Carbon dioxide (CO_2) is important because it contributes to the greenhouse effect, in which heat generated from sunlight at the Earth's surface is trapped by certain gasses and prevented from escaping through the atmosphere. The greenhouse effect itself is a perfectly natural phenomenon and, without it, the Earth would be a much colder place. But as is often the case, too much of a good thing can have negative consequences, and an unnatural buildup of greenhouse gasses can lead to a planet that gets unnaturally hot.

In recent years CO_2 has received much attention because its concentration in the atmosphere has risen to approximately 30% above natural background levels and will continue to rise into the near future. Scientists have shown that this increase is a result of human activities that have occurred over the last 150 years, including the burning of fossil fuels and deforestation. Because CO_2 is a greenhouse gas, this increase is believed to be causing a rise in global temperatures. This is the primary cause of climate change and is the main reason for increasing interest in the carbon cycle.

The Earth's carbon reservoirs naturally act as both sources, adding carbon to the atmosphere, and sinks, removing carbon from the atmosphere. If all sources are equal to all sinks, the carbon cycle can be said to be in equilibrium (or in balance) and there is no change in the size of the pools over time. Maintaining a steady amount of CO_2 in the atmosphere helps maintain stable average temperatures at the global scale. However, because fossil fuel combustion and deforestation have increased CO_2 inputs to the atmosphere without matching increases in the natural sinks that draw CO_2 out of the atmosphere (oceans, forests, etc.), these activities have caused the size of the atmospheric carbon pool to increase. This is what has been responsible for the present buildup of CO_2 and is believed to cause the observed trend of increasing global temperatures. How far will CO_2 levels rise in the future? The answer depends both on how much CO_2 humans continue to release and on the future amount of carbon uptake and storage by the Earth's natural sinks and reservoirs. In short, it depends on the carbon cycle.

1. We often refer to carbon occurring in "organic" versus "inorganic" forms. This is a simple way of grouping different forms of carbon into biologically derived compounds (complex substances produced only by the growth of living organisms) and mineral compounds that can be formed in the absence of biological activity (but can sometimes be formed with the assistance of living things, as in the case of sea shells). Organic compounds includes such things as sugars, fats, proteins and starches and are contained in both living organisms and the material that remains after their death and partial decomposition (including the organic matter in soils as well as the deposits of coal and

oils we refer to as fossil fuels). Note that complete decomposition of organic matter results in a return to mineral forms, often as CO_2. Mineral forms of carbon include carbonates contained in rock and seawater as well as CO_2 itself.

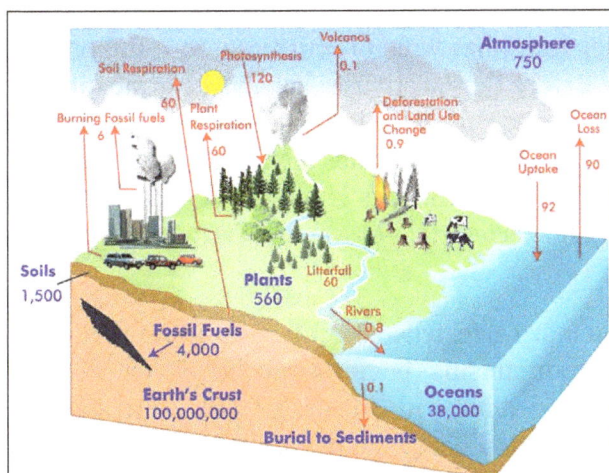

A simplified diagram of the global carbon cycle. Pool sizes, shown in blue, are given in petagrams (Petagram) of carbon. Fluxes, shown in red, are in Petagram per year.

Carbon Pools

The Earth's carbon pools can be grouped into any number of different categories. Here, we will consider four categories that have the greatest relevance to the overall carbon cycle.

1. The Earth's Crust: The largest amount of carbon on Earth is stored in sedimentary rocks within the planet's crust. These are rocks produced either by the hardening of mud (containing organic matter) into shale over geological time, or by the collection of calcium carbonate particles, from the shells and skeletons of marine organisms, into limestone and other carboncontaining sedimentary rocks. Together all sedimentary rocks on Earth store 100,000,000 Petagram Carbon (Petagrams of carbon). Recalling that 1 Petagram is is equal to a trillion kilograms and over two trillion pounds; this is clearly a large mass of carbon! Another 4,000 Petagram Carbon is stored in the Earth's crust as hydrocarbons formed over millions of years from ancient living organisms under intense temperature and pressure. These hydrocarbons are commonly known as fossil fuels.

2. Oceans: The Earth's oceans contain 38,000 Petagram Carbon, most of which is in the form of dissolved inorganic carbon stored at great depths where it resides for long periods of time. A much smaller amount of carbon, approximately 1,000 Petagram, is located near the ocean surface. This carbon is exchanged rapidly with the atmosphere through both physical processes, such as CO_2 gas dissolving into the water, and biological processes, such as the growth, death and decay of plankton. Although most of this surface carbon cycles rapidly, some of it can also be transferred by sinking to the deep ocean pool where it can be stored for a much longer time.

3. Atmosphere: The atmosphere contains approximately 750 Petagram Carbon, most of which is in the form of CO_2, with much smaller amounts of methane (CH_4) and various other compounds. Although this is considerably less carbon than that contained in the

oceans or crust, carbon in the atmosphere is of vital importance because of its influence on the greenhouse effect and climate. The relatively small size of the atmospheric C pool also makes it more sensitive to disruptions caused by an increase in sources or sinks of C from the Earth's other pools. In fact, the present-day value of 750 Petagram Carbon is substantially higher than that which occurred before the onset of fossil fuel combustion and deforestation. Before these activities began, the atmosphere contained approximately 560 Petagram Carbon and this value is believed to be the normal upper limit for the Earth under natural conditions. In the context of global pools and fluxes, the increase that has occurred in the past several centuries is the result of C fluxes to the atmosphere from the crust (fossil fuels) and terrestrial ecosystems (via deforestation and other forms of land clearing).

4. Terrestrial Ecosystems: Terrestrial ecosystems contain carbon in the form of plants, animals, soils and microorganisms (bacteria and fungi). Of these, plants and soils are by far the largest and, when dealing with the entire globe, the smaller pools are often ignored. Unlike the Earth's crust and oceans, most of the carbon in terrestrial ecosystems exists in organic forms. In this context, the term "organic" refers to compounds produced by living things, including leaves, wood, roots, dead plant material and the brown organic matter in soils (which is the decomposed remains of formerly living tissues).

Plants exchange carbon with the atmosphere relatively rapidly through photosynthesis, in which CO_2 is absorbed and converted into new plant tissues, and respiration, where some fraction of the previously captured CO_2 is released back to the atmosphere as a product of metabolism. Of the various kinds of tissues produced by plants, woody stems such as those produced by trees have the greatest ability to store large amounts of carbon, because wood is dense and trees can be large. Collectively, the Earth's plants store approximately 560 Petagram Carbon, with the wood in trees being the largest fraction.

The total amount of carbon in the world's soils is estimated to be 1500 Petagram Carbon. Measuring soil carbon can be challenging, but a few basic assumptions can make estimating it much easier. First, the most prevalent form of carbon in the soil is organic carbon derived from dead plant materials and microorganisms. Second, as soil depth increases the abundance of organic carbon decreases. Standard soil measurements are typically only taken to 1m in depth. In most cases, this captures the dominant fraction of carbon in soils, although some environments have very deep soils where this rule doesn't apply. Most of the carbon in soils enters in the form of dead plant matter that is broken down by microorganisms during decay. The decay process also releases carbon back to the atmosphere because the metabolism of these microorganisms eventually breaks most of the organic matter all the way down to CO_2.

Carbon Fluxes

The movement of any material from one place to another is called a flux and we typically think of a carbon flux as a transfer of carbon from one pool to another. Fluxes are usually expressed as a rate with units of an amount of some substance being transferred over a certain period of time (e.g. g cm^{-2} s^{-1} or kg km^2 yr^{-1}). For example, the flow of water in a river can be thought of as a flux that transfers water from the land to the sea and can be measured in liters per second, cubic meters per minute or cubic kilometers per year.

A single carbon pool can often have several fluxes both adding and removing carbon simultaneously. For example, the atmosphere has inflows from decomposition (CO_2 released by the breakdown of organic matter), forest fires and fossil fuel combustion and outflows from plant growth and uptake by the oceans. The size of various fluxes can vary widely.

1. Photosynthesis: During photosynthesis, plants use energy from sunlight to combine CO_2 from the atmosphere with water from the soil to create carbohydrates (notice that the two parts of the word, carbo- and –hydrate, signify carbon and water). In this way, CO_2 is removed from the atmosphere and stored in the structure of plants. Virtually all of the organic matter on Earth was initially formed through this process. Because some plants can live to be tens, hundreds or sometimes even thousands of years old (in the case of the longest-living trees), carbon may be stored, or sequestered, for relatively long periods of time. When plants die, their tissues remain for a wide range of time periods. Tissues such as leaves, which have a high quality for decomposer organisms, tend to decay quickly, while more resistant structures, such as wood can persist much longer. Current estimates suggest photosynthesis removes 120 Petagram Carbon/year from the atmosphere and about 610 Petagram Carbon is stored in plants at any given time.

2. Plant Respiration: Plants also release CO_2 back to the atmosphere through the process of respiration (the equivalent for plants of exhaling). Respiration occurs as plant cells use carbohydrates, made during photosynthesis, for energy. Plant respiration represents approximately half (60 Petagram Carbon/year) of the CO2 that is returned to the atmosphere in the terrestrial portion of the carbon cycle.

3. Litterfall: In addition to the death of whole plants, living plants also shed some portion of their leaves, roots and branches each year. Because all parts of the plant are made up of carbon, the loss of these parts to the ground is a transfer of carbon (a flux) from the plant to the soil. Dead plant material is often referred to as litter (leaf litter, branch litter, etc.) and once on the ground, all forms of litter will begin the process of decomposition.

4. Soil Respiration: The release of CO_2 through respiration is not unique to plants, but is something all organisms do, including microscopic organisms living in soil. When dead organic matter is broken down or decomposed (consumed by bacteria and fungi), CO_2 is released into the atmosphere at an average rate of about 60 Petagram Carbon/year globally. Because it can take years for a plant to decompose (or decades in the case of large trees), carbon is temporarily stored in the organic matter of soil.

5. Ocean—Atmosphere exchange: Inorganic carbon is absorbed and released at the interface of the oceans' surface and surrounding air, through the process of diffusion. It may not seem obvious that gasses can be dissolved into, or released from water, but this is what leads to the formation of bubbles that appear in a glass of water left to sit for a long enough period of time. The air contained in those bubbles includes CO_2 and this same process is the first step in the uptake of carbon by oceans. Once in a dissolved form, CO_2 goes on to react with water in what are known as the carbonate reactions. These are relatively simple chemical reactions in which H_2O and CO_2 join to form H_2CO_3 (also known as carbonic acid, the anion of which is called carbonate or CO_3). The formation of carbonate in seawater allows oceans to take up and store a much larger amount of carbon than would be possible if dissolved CO_2 remained in that form. Carbonate

is also important to a vast number of marine organisms that use this mineral form of carbon to build shells.

Carbon is also cycled through the ocean by the biological processes of photosynthesis, respiration, and decomposition of aquatic plants. In contrast with terrestrial vegetation is the speed at which marine organisms decompose. Because ocean plants don't have large, woody trunks that take years to breakdown, the process happens much more quickly in oceans than on land—often in a matter of days. For this reason, very little carbon is stored in the ocean through biological processes. The total amount of carbon uptake (92 Petagram C) and carbon loss (90 Petagram Carbon) from the ocean is dependent on the balance of organic and inorganic processes.

6. Fossil fuel combustion and land cover change: The carbon fluxes discussed thus far involve natural processes that have helped regulate the carbon cycle and atmospheric CO_2 levels for millions of years. However, the modern-day carbon cycle also includes several important fluxes that stem from human activities. The most important of these is combustion of fossil fuels: coal, oil and natural gas. These materials contain carbon that was captured by living organisms over periods of millions of years and has been stored in various places within the Earth's crust. However, since the onset of the industrial revolution, these fuels have been mined and combusted at increasing rates and have served as a primary source of the energy that drives modern industrial human civilization. Because the main byproduct of fossil fuel combustion is CO_2, these activities can be viewed in geological terms as a new and relatively rapid flux to the atmosphere of large amounts of carbon. At present, fossil fuel combustion represents a flux to the atmosphere of approximately 6-8 Petagram Carbon/year.

Another human activity that has caused a flux of carbon to the atmosphere is land cover change, largely in the form of deforestation. With the expansion of the human population and growth of human settlements, a considerable amount of the Earth's land surface has been converted from native ecosystems to farms and urban areas. Native forests in many areas have been cleared for timber or burned for conversion to farms and grasslands. Because forests and other native ecosystems generally contain more carbon (in both plant tissues and soils) than the cover types they have been replaced with, these changes have resulted in a net flux to the atmosphere of about 1.5 Petagram Carbon/year. In some areas, regrowth of forests from past land clearing activities can represent a sink of carbon (as in the case of forest growth following farm abandonment in eastern North America), but the net effect of all human-induced land cover conversions globally represents a source to the atmosphere.

7. Geological Processes: Geological processes represent an important control on the Earth's carbon cycle over time scales of hundreds of millions of years. A thorough discussion of the geological carbon cycle is beyond the scope of this introduction, but the processes involved include the formation of sedimentary rocks and their recycling via plate tectonics, weathering and volcanic eruptions.

To take a slightly closer look, rocks on land are broken down by the atmosphere, rain, and groundwater into small particles and dissolved materials, a process known as weathering. These materials are combined with plant and soil particles that result from decomposition and surface erosion and are later carried to the ocean where the larger particles are deposited near shore. Slowly, these sediments accumulate, burying older sediments below. The layering and burial of sediment causes pressure to build, which eventually becomes so great that deeper sediments are turned into rock, such as shale. Within the ocean water itself, dissolved materials mix with seawater and are used by marine life to make calcium carbonate ($CaCO_3$) skeletons and shells. When these organisms die, their skeletons and shells sink to the bottom of the ocean. In shallow waters (less than 4 km) the carbonate collects and eventually forms another type of sedimentary rock called limestone.

Collectively, these processes slowly convert carbon that was initially contained in living organisms into sedimentary rocks within the Earth's crust. Once there, these materials continue to be moved and transformed through the process of plate tectonics, uplift of rocks contained in the lighter plates and melting of rocks in the heavier plates as they are pushed deep under the surface. These melted materials can eventually result in emission of gaseous carbon back to the atmosphere through volcanic eruptions, thereby completing the cycle. Without this geological recycling, the carbon that becomes bound up in rocks would accumulate and remain there forever, eventually depleting the sources of CO_2 that are vital to life. The recycling of carbon through sedimentary rocks is an important part of our planet's long-term (over millions of years) ability to sustain life. Without it, the carbon that becomes bound up in rocks would accumulate and remain there forever, eventually depleting the sources of CO_2 that are vital to plants. However, because the geological cycle moves so slowly, these fluxes are small on an annual basis and have little effect on a human time-scale.

The Sulfur Cycle

Sulfur, an essential element for the macromolecules of living things, is released into the atmosphere by the burning of fossil fuels, such as coal. As a part of the amino acid cysteine, it is involved in the formation of disulfide bonds within proteins, which help to determine their 3-D folding patterns, and hence their functions. As shown in Figure, sulfur cycles between the oceans, land, and atmosphere. Atmospheric sulfur is found in the form of sulfur dioxide (SO_2) and enters the atmosphere in three ways: from the decomposition of organic molecules, from volcanic activity and geothermal vents, and from the burning of fossil fuels by humans.

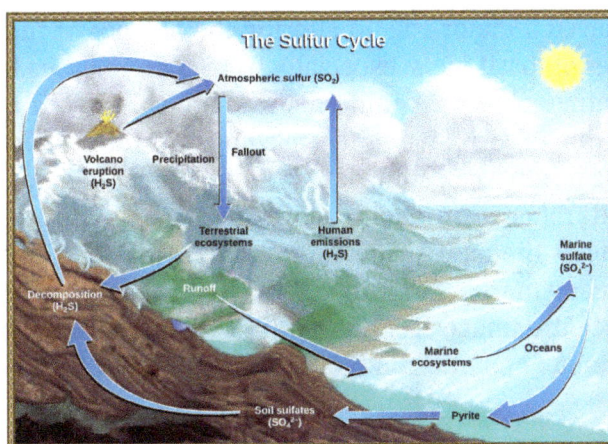

In figure, sulfur dioxide from the atmosphere becomes available to terrestrial and marine ecosystems when it is dissolved in precipitation as weak sulfuric acid or when it falls directly to the Earth as fallout. Weathering of rocks also makes sulfates available to terrestrial ecosystems. Decomposition of living organisms returns sulfates to the ocean, soil and atmosphere.

On land, sulfur is deposited in four major ways: precipitation, direct fallout from the atmosphere, rock weathering, and geothermal vents. Atmospheric sulfur is found in the form of sulfur dioxide (SO_2), and as rain falls through the atmosphere, sulfur is dissolved in the form of weak sulfuric acid (H_2SO_4). Sulfur can also fall directly from the atmosphere in a process called fallout. Also, the weathering of sulfur-containing rocks releases sulfur into the soil. These rocks originate from ocean sediments that are moved to land by the geologic uplifting of ocean sediments. Terrestrial ecosystems can then make use of these soil sulfates (SO_4^-), and upon the death and decomposition of these organisms, release the sulfur back into the atmosphere as hydrogen sulfide (H_2S) gas.

At this sulfur vent in Lassen Volcanic National Park in northeastern California, the yellowish sulfur deposits are visible near the mouth of the vent.

Sulfur enters the ocean via runoff from land, from atmospheric fallout, and from underwater geothermal vents. Some ecosystems rely on chemoautotrophs using sulfur as a biological energy source. This sulfur then supports marine ecosystems in the form of sulfates.

Human activities have played a major role in altering the balance of the global sulfur cycle. The burning of large quantities of fossil fuels, especially from coal, releases larger amounts of hydrogen sulfide gas into the atmosphere. As rain falls through this gas, it creates the phenomenon known as acid rain. Acid rain is corrosive rain caused by rainwater falling to the ground through sulfur dioxide gas, turning it into weak sulfuric acid, which causes damage to aquatic ecosystems. Acid rain damages the natural environment by lowering the pH of lakes, which kills many of the resident fauna; it also affects the man-made environment through the chemical degradation of buildings. For example, many marble monuments, such as the Lincoln Memorial in Washington, DC, have suffered significant damage from acid rain over the years. These examples show the wide-ranging effects of human activities on our environment and the challenges that remain for our future.

The Oxygen Cycle

Oxygen cycle along with the carbon cycle and nitrogen cycle plays an important role in the existence of life on the earth. The oxygen cycle is a biological process which helps in maintaining the oxygen level by moving through three main regions of the earth which are:

1. Atmosphere

2. Lithosphere

3. Biosphere

This biogeochemical cycle explains the movement of oxygen gas within the atmosphere, the ecosystem, biosphere and the lithosphere. The oxygen cycle is interconnected with the carbon cycle.

The atmosphere is the layer of gases presents above the earth's surface. The sum of all Earth's ecosystem makes a biosphere. Lithosphere, which is the solid outer section along with the Earth's crust and it is the largest reservoir of oxygen.

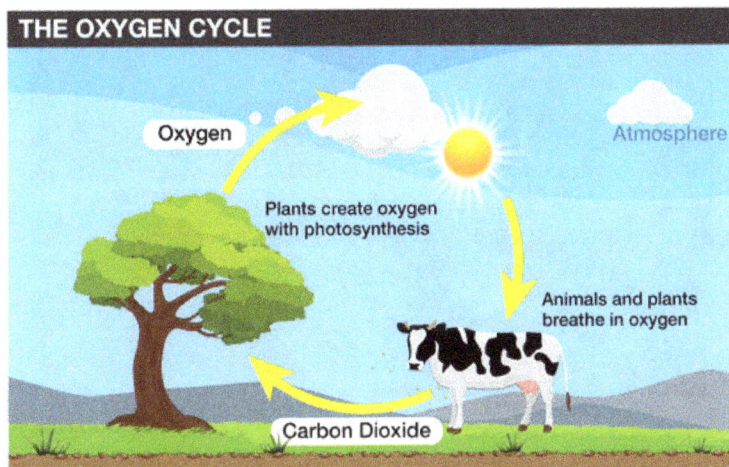

Stages of the Oxygen Cycle

The steps involved in the oxygen cycle are:

Stage-1: All green plants during the process of photosynthesis, release oxygen back into the atmosphere as a by-product.

Stage-2: All aerobic organisms, uses free oxygen for respiration.

Stage-3: Animals exhale Carbon dioxide back into the atmosphere which is again used by the plants during photosynthesis. Now oxygen is balanced within the atmosphere.

Uses of Oxygen

The four main processes that use Atmospheric oxygen are:

1. Breathing – It is the physical process, through which all living organisms including plants, animals, and humans inhale oxygen from the outside environment into the cells of an organism and exhale carbon dioxide back into the atmosphere.

2. Decomposition: It is one of the natural and most important processes in the oxygen cycle and occurs when an organism dies. The dead animal or plants decay into the ground and the organic matter along with the carbon, oxygen, water and other components are returned back into the soil and air. This process is carried out by the invertebrates including fungi, bacteria and some insects which are collectively called as the decomposers. The entire process requires oxygen and releases carbon dioxide.

3. Combustion: It is also one of the most important processes which occur when any of the organic materials including fossil fuels, plastics and wood, are burned in the presence of oxygen and releases carbon dioxide into the atmosphere.

4. Rusting: This process also requires oxygen. It is the formation of oxides which is also called oxidation. In this process, metals like iron or alloy rust when they are exposed to moisture and oxygen for a long period of time and new compounds of oxides are formed by the combination of oxygen with the metal.

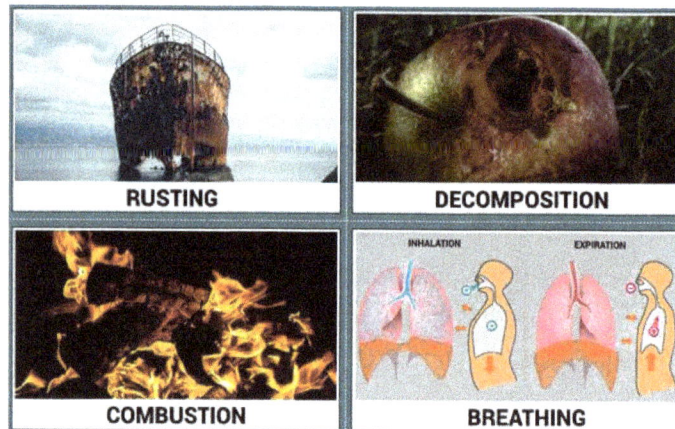

Production of Oxygen

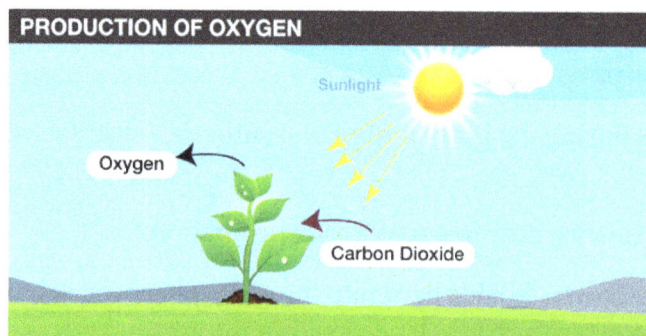

1. Plants: The main creators of oxygen are plants by the process of photosynthesis. Photosynthesis is a biological process by which all green plants synthesize their food in the presence of sunlight. During photosynthesis, plants use sunlight, water, carbon dioxide to create energy and oxygen gas is liberated as a by-product of this process.

2. Sunlight: Sunlight also produces oxygen. Some oxygen gas is produced when the sunlight reacts with water vapour in the atmosphere.

Some Interesting Facts About Oxygen

- Phytoplankton is one of the biggest sources of oxygen, followed by terrestrial plants and trees.

- Oxygen is also produced when the sunlight reacts with water vapour present in the atmosphere.

- A large amount of oxygen is stored in the earth's crust in the form of oxides, which cannot be used for respiration process as it is available in the combined state.

Importance of Oxygen Cycle

As we all know, Oxygen is one of the most important components of the Earth's atmosphere. It is mainly required for:

- Breathing,

- Combustion,

- Supports aquatic life,

- Decomposition of organic waste,

- Lack of bottom oxygen is the major cause of odours and acids produced by the anaerobic bacteria.

The oxygen cycle is mainly involved in maintaining the level of Oxygen in the atmosphere. The entire cycle can be summarized as, the oxygen cycle begins with the process of photosynthesis in the presence of sunlight, releases oxygen back into the atmosphere, which humans and animals breathe in oxygen and breathe out carbon dioxide, and again linking back to the plants. This also proves that both the oxygen and carbon cycle occur independently and are interconnected to each other.

Geochemistry

Geochemists study the chemical composition of Earth and other planets.

The field of geochemistry involves study of the chemical composition of Earth and extraterrestrial bodies and systems, and the chemical processes and reactions that take place within them. It also involves investigation of the cycles of matter and energy that transport the Earth's chemical constituents through time and space.

Scientific studies in geochemistry provide knowledge about Earth and its history, and they help us understand some of the processes involved in the formation of valuable mineral deposits and in changing the planet's climate. Geochemical knowledge is also useful when making plans to dispose of toxic wastes in a manner that causes least harm to humans and the environment.

Subfields

Geochemistry includes the following major subfields and areas of study:

- Cosmochemistry: It deals with analysis of the distribution of elements and their isotopes in extraterrestrial bodies and systems. Studies in cosmochemistry include attempts to understand the formation of and chemical processes within the Solar System, the origin of meteorites, and the formation of elements in stars.

- Examination of the distribution and movements of elements in different parts of Earth (the crust, mantle, hydrosphere, and so forth) and in minerals, with the goal of determining the underlying systems of distribution and transport.

- Isotope geochemistry: It involves determining the distribution and concentrations of the isotopes of elements in terrestrial and extraterrestrial materials. The knowledge gained may be used to determine the age of these materials and the historical changes they have gone through.

- Organic geochemistry: This area involves studying the role of carbon-containing compounds and processes derived from living or once-living organisms. This area of geochemistry helps us understand how living things affect chemical cycles, and the formation of petroleum, coal, natural gas, and ores.

- Regional, environmental and exploration geochemistry: It involves studies related to environmental, hydrological, and mineral exploration.

Meteorites may be studied as part of cosmochemistry.

Chemical Characteristics of Rocks

The more common constituents of rocks on Earth are oxides. The main exceptions to oxides are compounds of chlorine, sulfur, and fluorine.

According to calculations by F. W. Clarke, a little more than 47 percent of Earth's crust consists of oxygen. It occurs mainly in the form of oxides, particularly silica, alumina, iron oxides, lime, magnesia, potash, and soda. Silica functions principally as an acid, forming silicates, and the most common minerals of igneous rocks are silicates. From a computation based on 1,672 analyses of all kinds of rocks, Clarke arrived at the following values for the average percentage composition: SiO_2=59.71; Al_2O_3=15.41; Fe_2O_3=2.63; FeO=3.52; MgO=4.36; CaO=4.90; Na_2O=3.55; K_2O=2.80; H_2O=1.52; TiO_2=0.60; and P_2O_5=0.22. (The total of these is 99.22 percent). All other constituents occur in very small quantities, usually much less than one percent.

The oxides combine in various ways. Some examples are given below:

- Potash and soda combine to produce mostly feldspars, but may also produce nepheline, leucite, and muscovite.

- Phosphoric acid with lime forms apatite.

- Titanium dioxide with ferrous oxide gives rise to ilmenite.

- Magnesia and iron oxides with silica crystallize as olivine or enstatite, or with alumina and lime form the complex ferro-magnesian silicates (such as the pyroxenes, amphiboles, and biotites).

- Any silica in excess of that required to neutralize the bases separates out as quartz; excess alumina crystallizes as corundum.

These combinations must be regarded only as general tendencies, for there are numerous exceptions to the rules. The prevalent physical conditions also play a role in the formation of rocks.

Clarke also calculated the relative abundances of the principal rock-forming minerals and obtained the following results: apatite=0.6 percent, titanium minerals=1.5 percent, quartz=12.0 percent, feldspars=59.5 percent, biotite=3.8 percent, hornblende and pyroxene=16.8 percent, for a total of 94.2 percent. These figures, however, can only be considered rough approximations.

Acid, Intermediate, Basic and Ultrabasic Igneous Rocks

Rocks that contain the highest levels of silica and on crystallization yield free quartz are placed in a group generally designated "acid" rocks. Rocks that contain lowest levels of silica and most magnesia and iron, so that quartz is absent while olivine is usually abundant, form the "basic" group. The "intermediate" group includes rocks characterized by the general absence of both quartz and olivine. An important subdivision of these contains a very high percentage of alkalis, especially soda, and consequently has minerals such as nepheline and leucite not common in other rocks. It is often separated from the others as the "alkali" or "soda" rocks, and there is a corresponding series of basic rocks. Lastly, a small group rich in olivine and without feldspar has been called "ultrabasic" rocks. They have very low percentages of silica but high proportions of iron and magnesia.

Except for the last group, practically all rocks contain feldspars or feldspathoid minerals. In acid rocks, the common feldspars are orthoclase, with perthite, microcline, oligoclase, all having much silica. In the basic rocks, labradorite, anorthite, and bytownite prevail, being rich in lime and poor in silica, potash and soda. Augite is the most common ferro-magnesian of the basic rocks, while biotite and hornblende are usually more frequent in acid rocks.

	Acid	Intermediate		Basic	Ultrabasic
Commonest Minerals	Quartz Orthoclase (and Oligoclase), Mica, Hornblende, Augite	Little or no Quartz: Orthoclase hornblende, Augite, Biotite	Little or no Quartz: Plagioclase Hornblende, Augite, Biotite	No Quartz Plagioclase Augite, Olivine	No Felspar Augite, Hornblende, Olivine
Plutonic or Abyssal type	Granite	Syenite	Diorite	Gabbro	Peridotite
Intrusive or Hypabyssal type	Quartz-porphyry	Ortho-clase-porphyry	Porphyrite	Dolerite	Picrite
Lavas or Effusive type	Rhyolite, Obsidian	Trachyte	Andesite	Basalt	Limburgite

Rocks that contain leucite or nepheline, either partly or wholly replacing feldspar, are not included in the above table. They are essentially of intermediate or basic character. They may be regarded as varieties of syenite, diorite, gabbro, and so forth, in which feldspathoid minerals occur. Indeed there are many transitions between ordinary syenites and nepheline (or leucite) syenite, and between gabbro or dolerite and theralite or essexite. But because many minerals that develop in these "alkali" rocks are uncommon elsewhere, it is convenient in a purely formal classification like that outlined here to treat the whole assemblage as a distinct series.

Nepheline and Leucite-bearing Rocks			
Commonest Minerals	Alkali Feldspar, Nepheline or Leucite, Augite, Hornblend, Biotite	Soda Lime Feldspar, Nepheline or Leucite, Augite, Hornblende (Olivine)	Nepheline or Leucite, Augite, Hornblende, Olivine
Plutonic type	Nepheline-syenite, Leucite-syenite, Nepheline-porphyry	Essexite and Theralite	Ijolite and Missourite
Effusive type or Lavas	Phonolite, Leucitophyre	Tephrite and Basanite	Nepheline-basalt, Leucite-basalt

The above classification is based essentially on the mineralogical constitution of igneous rocks. Any chemical distinctions between the different groups, though implied, are relegated to a subordinate position. It is admittedly artificial, but it has developed with the growth of the science and is still adopted as the basis on which smaller subdivisions have been set up.

The subdivisions are by no means of equal value. For example, the syenites and the peridotites are far less important than the granites, diorites, and gabbros. Moreover, the effusive andesites do not always correspond to the plutonic diorites but partly also to the gabbros.

As the different types of rock, regarded as aggregates of minerals, pass gradually from one to another, transitional types are very common and are often so important as to receive special names.

For example, the quartz-syenites and nordmarkites may be interposed between granite and syenite, the tonalites and adamellites between granite and diorite, the monzoaites between syenite and diorite, and the norites and hyperites between diorite and gabbro.

Geochemical Distribution of the Elements

Terrestrial Distribution

The study of earthquake waves passing through the body of the Earth has shown that the interior is not uniform; it consists of distinct shells separated by concentric discontinuities at which the velocities of the passing waves change. The two major discontinuities that are universally recognized are the Mohorovičić Discontinuity, which divides the Earth's crust from its underlying mantle, and the Wiechert–Gutenberg Discontinuity, which separates the mantle from the core. The latter discontinuity exists at a depth of 2,900 kilometres (1,800 miles); it is marked by a sudden increase in density, from about 5.7 at the base of the mantle to 9.7 at the top of the core. The only reasonable interpretation of this discontinuity is that the mantle consists of silicates and oxides of the common elements (largely magnesium and iron), and the core consists of metallic iron alloyed with minor amounts of other elements (analogous to the nickel-iron in meteorites). The Mohorovičić Discontinuity varies in depth from place to place; it averages about 33 kilometres (20 miles) below the continents and about 8 kilometres (5 miles) below the bottom of the deep oceans. It too is marked by a density increase from crust to mantle—a comparatively small one, from about 3 to 3.3.

To the three spherical divisions—crust, mantle, and core—two more should be added: the hydrosphere, which is the discontinuous shell of fresh and salt water, on and within the crust; and the atmosphere, the ocean of air that surrounds the Earth, gradually thinning into the vacuum of outer space.

The Earth's Core

The evidence for the composition of the core is all indirect because no means have yet been devised for directly sampling the deep interior of the Earth. The moment of inertia of the Earth indicates that there is a concentration of mass around the centre, and seismic data have shown that below the Wiechert–Gutenberg Discontinuity the density of the material is high, ranging upwards from 9.7. The only heavy element with high cosmic abundance is iron, and because an iron–nickel alloy is an important meteorite component, it is reasonable to conclude that the Earth's core consists largely of metallic iron with a minor admixture of other elements. This conclusion is supported by geophysical evidence that indicates that the mean atomic number of the material of the core is about 22. The atomic number of iron is 26, so this implies that the core also contains elements of lower atomic number. Sulfur, with atomic number 16, and carbon, 6, are relatively abundant in meteoritic matter, and the presence of minor amounts of these elements in the core would effectively reduce the mean atomic number. Some authorities have advocated silicon (atomic number 14) as the major alloying component in the core, but this seems less likely; if silicon were the sole alloying element, then the core would have to contain more than 30 percent silicon in order to reduce its mean atomic number to 22. In addition, free silicon requires extremely reducing conditions (lack of oxygen), and the presence of ferrous iron in the mantle is inconsistent with this requirement.

It is not possible to give definite figures for the abundances of the elements in the Earth's core. It is certainly made up largely of metallic iron, however, probably with some nickel, a little cobalt, and appreciable amounts of such lighter elements as carbon and sulfur.

The Earth's Mantle

The mantle comprises that part of the Earth between the Mohorovičić and the Wiechert–Gutenberg discontinuities. It makes up 83 percent of the volume of the Earth and 67 percent of its mass and is thus of decisive importance in determining the bulk composition of the planet. In estimating elemental abundances in the mantle, however, the same difficulty as with the core arises: direct sampling is not feasible. Much more geophysical data are available for the mantle, however, and some volcanic eruptions have brought rock fragments to the surfaces that have certainly been derived from this zone. The most remarkable of these materials are the diamond-bearinginclusions found in the famous pipes, or volcanic necks, that are mined in South Africa and Siberia. The presence of diamond, the high-pressure form of carbon, implies a depth of origin of at least 100 kilometres (62 miles), but these inclusions are rare. The common type of mantle-derived inclusion is peridotite, a silicate rock consisting largely of olivine, $(Mg,Fe)_2SiO_4$, with minor amounts of orthopyroxene, $(Mg,Fe)SiO_3$, and diopside, $CaMg(Si_2O_6)$.

Geophysical information indicates that below a depth of about 1,000 kilometres (620 miles), the mantle behaves as an essentially homogeneous material, but above this level its physical properties are more varied, and there is evidence for second-order discontinuities. This region above 1,000 kilometres is frequently referred to as the upper mantle, and in recent years has been the object of a concentrated research effort by geologists and geophysicists all over the world. The significance of the upper mantle is that processes originating there have dramatic effects on the surface—in the form of volcanic eruptions and some earthquakes—and less dramatic but equally important effects within the crust, such as the introduction and concentration of some elements, possibly leading to the formation of ore deposits. Increased knowledge of the upper mantle thus has both scientific and economic appeal.

Geophysical data on the properties of the upper mantle suggest that it must consist essentially of magnesium-iron silicates, probably largely olivine in the region immediately below the crust. Olivine is not stable under very high pressures, however; it is converted to a different phase of about 10 percent higher density and with a structure like the mineral oxide spinel $(MgA1_2O_4)$. This conversion would occur in the mantle at depths of around 400 kilometres, and a second-order discontinuity at that depth can plausibly be ascribed to this conversion. Pyroxenes also undergo transformations to phases of greater density at the high pressures within the mantle. Thus the mantle, although composed of material of familiar chemical composition, consists, in its lower part at least, of different minerals than those in the upper part.

Many estimates of the composition of the upper mantle have been made in recent years. On the whole, the similarities are more important than the differences. All agree that the principal components are oxides of silicon, magnesium, and iron. The differences are mainly in the minor components such as aluminum oxide, calcium oxide, and the alkalies, and are determined largely by theoretical considerations and the weight given to specific aspects of the geophysical and geochemical data.

Although fairly reliable estimates exist for the abundances of the major elements in the mantle, little is known of minor and trace elements. Knowledge of the crystal structure of possible mantle minerals indicates that many minor and trace elements will not be readily incorporated, however. They are therefore likely to concentrate in liquid material in the mantle and be carried upward in solution, eventually being transported into the crust. It is thus probable that the mantle is relatively depleted, and the crust relatively enriched, in minor and trace elements. This is certainly true for uranium and thorium, because the amount of these elements in the crust is almost sufficient to account for the total amount of heat flowing out of the Earth.

The Earth's Crust

The crust is a comparatively thin shell on the surface of the Earth and makes up less than 1 percent of its total mass. Its geochemical significance is only marginally related to its bulk, however. It has been subjected to extensive investigations, and it provides the raw materials on which civilization depends. It is the most diverse of the geospheres, being a complex mosaic of many rock types—igneous, sedimentary, and metamorphic—each with a wide variety of chemical and mineralogical compositions. The surface is veneered with soils, related in composition to the rocks from which they formed, but with important modifications because of the smaller grain size, the presence of organic matter, and an intricate complex of living organisms. Ultimately, man's welfare and indeed his survival depend on the wise utilization of the materials in the crust. Modern civilization has been erected upon the exploitation of fuels and ore deposits, which are simply geochemical concentrations of useful elements.

Igneous Rocks

Clarke estimated that 95 percent of crustal rocks are of igneous origin (formed from molten silicate masses, or magmas). Sedimentary rocks occur as a thin veneer on an igneous or metamorphic basement, except where locally thickened in mountain belts. The primordial rocks of the crust must have been essentially igneous, and the first sedimentary rocks were derived from them by processes of weathering and erosion. Metamorphic rocks are formed from both sedimentary and igneous rocks by transformations due to heat and pressure at depth in the crust; unless very intense; these transformations do not totally obliterate the primary igneous or sedimentary features.

Major Components

Igneous rocks show a wide range of composition; the principal component, silica (SiO_2), ranges from about 35 percent to 80 percent among the commoner igneous rocks, and other components also show a wide variation. They thus illustrate some quite extensive geochemical fractionations of the elements, the fractionations that may have economic significance if they bring about the findings of workable ore deposits.

The average of igneous rock analyses from the oceanic islands is notably lower in silica and alkalies, and higher in magnesium and calcium oxides, than the continental averages. This is simply a reflection of the fact that most oceanic islands, such as Hawaii, consist almost entirely of basalts (averaging about 50 percent silica), whereas continental areas include large granitic masses, with silica contents around 70 percent. In terms of volumes, igneous rocks consist predominantly of two great types, granitic and basaltic. The former essentially are confined to the continents and the latter occur in both continents and ocean basins. The other types of igneous rocks, while many and varied, are quantitatively insignificant and hardly affect the averages. Thus, for the major elements, the average of over 5,000 analyses of igneous rocks is not significantly different from the simple average of two individual rocks, granite (G-1) and a basalt (W-1). This can be seen from the Table, by comparing G-1 and W-1 values with those in the column headed "Earth's crust."

The figures for the specific granitic (G-1) and basaltic (W-1) rocks are included in the Table because they have been analyzed for practically all the elements in different geochemical laboratories throughout the world. The rock G-1 was a granite from Westerly, Rhode Island, and W-1 was a basaltic rock (specifically a diabase) from Centerville, Virginia. Several hundred kilograms of each of these rocks were crushed to a fine powder in the laboratories of the U.S. Geological Survey and samples distributed to analytical laboratories throughout the world, in order to obtain as many analyses as possible. G-1 and W-1 are undoubtedly the most thoroughly analyzed rocks and now serve as basic geochemical standards. It must be borne in mind that they are individual rocks, however, and cannot be considered to be averages of all granites and all basalts. Indeed, it is clear that G-1 is unusually rich in some trace elements (e.g., thorium).

Perhaps the most significant feature of the composition of the Earth's crust is that it is dominated by comparatively few elements. Only eight—oxygen, silicon, aluminum, iron, calcium, magnesium, sodium, and potassium—are present in amounts greater than 1 percent, and these eight make up almost 99 percent of the whole. Of these, oxygencomprises almost 50 percent by weight. The dominance of oxygen is even more marked when weight percentages are converted to atomic percentages, as follows: oxygen (weight percentage, 46.6; atomic percentage, 62.2), silicon (27.7; 21.2), aluminum (8.13; 6.47), iron (5.00; 1.92), calcium (3.63; 1.94), magnesium (2.09; 1.84), sodium (2.83; 2.64), potassium (2.59; 1.42). This comparison, of course, merely emphasizes the fact that the crust consists almost entirely of oxygen compounds, mostly silicates and aluminosilicates of iron, calcium, magnesium, and the alkali metals. As Goldschmidt remarked, the lithosphere may well be called the oxysphere. Clarke and his collaborators calculated that the average mineralogical composition of igneous rocks is: quartz 12.0 percent, feldspars 59.5 percent, pyroxene and hornblende 16.8 percent; biotite 3.8 percent, titanium minerals 1.5 percent, apatite 0.6 percent, and other accessory minerals 5.8 percent.

Elements of Minor and Trace Abundance

A cursory examination immediately reveals some intriguing features in the abundance pattern. The predominance of the even-numbered elements over the neighbouring odd-numbered ones is still apparent but not so regular as in the cosmic abundances. Chemical fractionations taking place during the evolution of igneous rocks from primordial matter have clearly modified this basic relationship. The odd–even relationship is most prominent in the rare-earth elements (atomic numbers 57–71), which are chemically so similar that they are little fractionated by geochemical processes.

A particularly noteworthy feature is that some unfamiliar elements, such as rubidium, are relatively abundant, whereas others, such as most of the industrial metals except iron and aluminum, are actually of very low abundance. Thus boron, familiar to every homemaker in the form of borax cleansers and boric-acid antiseptics and well-known in the ancient world, is an element of extremely low abundance, much lower.

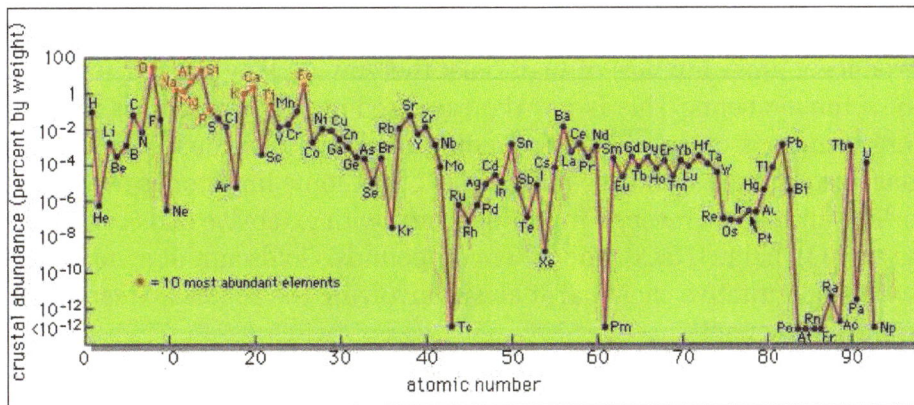

Crustal abundances of elements of atomic numbers 1 to 93.

The distribution of minor and trace elements in igneous rocks, for those of lithophile affinity, is largely controlled by their ionic radii or size. Minor and trace elements with radii similar to those of major elements can substitute for these elements in the common minerals of the igneous rocks. The crystal structures of these minerals act as sorting mechanisms, accepting those atoms of appropriate size and rejecting others. Thus rubidium, with a radius of 1.47Å (one angstrom [Å] = 10^{-8} centimetres) is incorporated in potassium feldspar, $KAlSi_3O_8$, because its radius is close to that of potassium (1.33Å). The next higher alkali element, cesium, with a considerably larger radius (1.67Å), is not accepted into the feldspar structure; it remains in the igneous liquid during the crystallization of the major minerals until its concentration increases to such an extent that it can form the independent mineral pollucite ($CsAlSi_2O_6$).

The factor of ionic size, coupled with geochemical affinity (lithophile, chalcophile, or siderophile) is a key to the distinction between abundance and availability. Elements that are similar in size and geochemical affinity to major elements are dispersed in small amounts in common minerals; i.e., rubidium in potassium feldspar, gallium in aluminum minerals, and germanium in silicate minerals. Elements that do not readily enter the common minerals of igneous rocks remain in the residual melt as crystallization proceeds. Fractional crystallization of magmas (igneous melts) normally results in a residual liquid of granitic composition. Under suitable conditions this residual liquid solidifies as a coarse-grained rock known as a pegmatite. Pegmatites are famous for their content of rare and unusual minerals, which contain many of the minor and trace elements. They are the commercial sources of lithium, beryllium, scandium, yttrium, the rare earths, cesium, niobium, and tantalum, all elements that concentrate in the residual liquid because of their specific geochemical properties.

Chalcophile elements are all of rather low abundance, and the minerals that they form, mainly sulfides and some arsenides, are not stable at the high temperatures of igneous crystallization. Sometimes these elements are found in granites and pegmatites—molybdenite (MoS_2) is a typical example. More frequently they are removed from the crystallizing magma as hot aqueous solutions and are deposited as metalliferous veins in the surrounding rock. Sometimes they may reach the surface as components in thermal springs; mercury has been deposited (as native mercury and as cinnabar, HgS) by some of these springs, occasionally in sufficient amounts for profitable mining.

Sedimentary Rocks

The decomposition of pre-existing rocks by weathering, the transportation and deposition of the weathering products as sediments, and the eventual formation of sedimentary rocks might be expected to produce a gross mixture of materials, thereby working against further geochemical differentiation of the elements. This is not the case; sedimentary processes frequently produce remarkable concentrations of the elements, leading to almost pure deposits of certain minerals. Some sandstone, e.g., contain over 99 percent quartz and some limestones over 99 percent calcium carbonate. The ultimate is reached in salt deposits, with extensive beds of anhydrite ($CaSO_4$), gypsum ($CaSO_4 \cdot 2H_2O$), halite ($NaCl$), and other compounds. Goldschmidt compared the sedimentary process with a quantitative chemical analysis, involving the successive separation of specific elements or groups of elements.

Quartz (SiO_2) is highly resistant to weathering and accumulates as deposits of sand. When consolidated these deposits form sandstones, an important group of sedimentary rocks. Under special conditions almost any mineral may be deposited in sand-sized grains, but most minerals are eventually decomposed by weathering. A few resistant ones may survive and be sufficiently concentrated to form economic deposits known as placers; the most familiar are probably the gold-bearing sands, important sources of this element, but sand deposits may have economic concentrations of zirconium (as the mineral zircon, $ZrSiO_4$), titanium (as rutile, TiO_2, and ilmenite, $FeTiO_3$), tin (as cassiterite, SnO_2), and others.

The aluminosilicates of igneous rocks, mainly the feldspars, $(K,Na)AlSi_3O_8$ and $(Na,Ca)(Al,Si)_4O_8$, are relatively easily decomposed by weathering. The alkali elements and calcium are largely carried away in solution, whereas the aluminum and silicon are quickly redeposited as insoluble clay minerals. When consolidated, these minerals form shales and mudstones. The ferromagnesian minerals undergo a more complex decomposition, sometimes leading to the deposition of iron-rich sediments consisting largely of hydrated ferric oxide; such sediments are valuable iron ores in many countries.

Calcium is carried away in solution mainly as calcium bicarbonate, $Ca(HCO_3)_2$. Most of it eventually reaches the sea, where it is utilized by a vast variety of organisms as skeletal material in the form of calcite and aragonite (polymorphs—different forms—of $CaCO_3$). Accumulation of skeletal materials after death of the organisms has formed extensive deposits of limestone throughout geological time. Magnesium in seawater can react with calcium carbonate to form dolomite, $CaMg(CO_3)_2$, and in this way some magnesium is removed from solution and deposited in sediments.

Much of the magnesium, however, remains in seawater, which is essentially a dilute solution of magnesium, calcium, sodium, and potassium chlorides and sulfates, with many other elements in small amounts. Under special geological circumstances bodies of seawater can be cut off from the open ocean, and under arid conditions the water will evaporate and extensive salt beds be deposited. Such conditions have occurred in different regions throughout geological time, and the resulting salt deposits are economically important as sources of sodium, potassium, calcium, magnesium, chlorine, and sulfur.

The three major groups of sedimentary rocks are sandstones, shales, and the carbonate rocks (limestones and dolomites). Much less geochemical research has been devoted to sedimentary rocks than to igneous rocks, and the data for their contents of minor and trace elements are therefore less extensive.

The problem of arriving at an average composition for all sedimentary rocks is still largely unresolved, largely because of uncertainty in the relative amounts of shales, sandstones, and carbonate rocks. From geochemical arguments Clarke estimated the relative percentages of these three groups as 80:15:5, respectively. Actual measurements of sedimentary rocks suggest that these figures overestimate the amount of shales and underestimate that of limestones, however. Thus, a compilation of the recorded amounts of shales, sandstones, and limestones in more than 213,000 metres (700,000 feet) of sedimentary rock formations gave relative percentages of 46:32:22, respectively. The identification of a formation as a limestone, a sandstone, or a shale, however, is likely to be gross; shales usually contain considerable sand, sandstones may carry much clay, and the term limestone is applied to many rocks with 50 percent or less of carbonate. It does appear that limestones are more prominent in the geological record than might be expected from geochemical calculations, however; this probably reflects the fact that shallow-water environments are the great places of carbonate deposition, whereas the ocean deeps are the repository primarily of clay-rich sediments.

Metamorphic Rocks

Comparatively few investigations have been made of the elemental composition of metamorphic rocks. Many of these rocks retain the geochemical features of their parent igneous or sedimentary materials, and their bulk composition has been little changed despite complete recrystallization and the production of new minerals and structures in some instances. Some metamorphic rocks, however, have been markedly modified by the removal of some components and the addition of others.

The Geological Survey of Canada has performed a comprehensive study of a large area of the Canadian Shield, a region of complex geology largely made up of metamorphic rocks. From a collection of more than 8,000 bedrock samples, the average abundances of all the major elements and a number of minor and trace elements were determined; the figures are given in the table. As might be anticipated, the average composition is not very different from the average composition of igneous rocks. It does show somewhat higher silicon content, probably reflecting a preponderance of granitic over basaltic rocks and a relative abundance of quartz-rich sedimentary rocks in the original makeup of the Canadian Shield. The general validity of these abundance figures for metamorphic rocks has been confirmed by a similar study of the average composition of metamorphic rocks in the former Soviet Union, which has given closely comparable results.

Ore Deposits

An ore deposit, in its simplest terms, is a portion of the Earth's crust from which some industrial raw material can be extracted at a profit. As such, its characteristics are as much economic as geochemical. Nevertheless, its formation required the operation of geochemical processes to produce the concentration of a specific element or elements in a particular place. Economics decide whether this concentration is rated as an ore deposit or merely as a deposit of scientific interest.

The economics may change with time, depending upon price, availability of transportation, cost of labour, and other factors.

Some general principles can, however, be enunciated. Proceeding from the average abundance of an element in the crust, and the minimum abundance that can be profitably exploited under normal circumstances, a factor of enrichment necessary to produce an ore deposit can be derived. The economic control is immediately evident in the approximate relation between the factor of enrichment and the price of the product sought. The most extreme example of this is in diamond mining, where the product sought may be present in the rock mined in as low a concentration as 1 part in 50,000,000. Ease of extraction, of course, plays an important role in this. Diamonds are readily separated from the great mass of waste rock by a relatively simple and inexpensive process. Magnesium is commercially extracted from seawater, where its concentration is 0.13 percent, rather than from the common rock dunite, where its concentration is about 25 percent, because of the ready availability of seawater and the cheapness of the extraction process.

Concentration factors for ore bodies of common metals			
Metal	Percent in earth's crust	Minimum percent profitably extracted	Enrichment factor necessary for an ore body
Aluminum	8.13	30	1
Iron	5.00	30	6
Manganese	0.10	35	350
Chromium	0.02	30	1,500
Copper	0.007	1	140
Nickel	0.008	1.5	175
Zinc	0.013	4	300
Tin	0.004	1	250
Lead	0.0016	4	2,500
Uranium	0.0002	0.1	500

Ore deposits may be found in all types of rocks—igneous, sedimentary, and metamorphic—and seawater is also a significant source of such elements as sodium, chlorine, magnesium, and bromine. There are many processes of geochemical enrichment leading to the formation of ore deposits, and they are often the end result of a complex series of such processes acting over a long period of time. The economic importance of ore deposits has ensured their thorough study by all techniques of geological and geochemical research, but much controversy still exists regarding the origin of many of the more complex deposits.

The most readily understood ore deposits are those of sedimentary origin. They have been formed at the surface of the Earth by processes that can usually be observed operating at the present time and that can readily be simulated in the laboratory. Salt deposits are one kind whose origin is clearly amenable to such an approach. As long ago as 1849 an Italian scientist initiated laboratory studies on the evaporation of seawater and elucidated the sequence of crystallization of the different salts. Comparison of the results with the mineralogy of salt deposits revealed gross similarities but also important differences; these differences can be explained by a variety of mild metamorphic reactions resulting from burial of these deposits under overlying sediments.

Some sedimentary deposits are not readily explicable by such an approach, however. The most extensive and economically important are the vast Precambrian iron ore deposits, which are a major source for the hundreds of millions of tons of steel produced annually. They occur on all the continents (except perhaps Antarctica) and are uniformly of great age (about 1,900,000,000 years or older). Probably the most extensive and best exposed of these are in the Hamersley Range of Western Australia, where individual beds of iron ore are continuous over hundreds of square miles in a horizontally bedded sequence of iron ore and quartzite thousands of feet thick. The conditions that gave rise to these deposits were apparently unique to this early period in Earth history, because similar deposits are not known in younger geological formations. It has been argued that the explanation lies in an oxygen-free reducing atmosphere in early geological times, under which iron could readily be transported in solution as ferrous compounds to the ocean or large lakes, where deposition eventually took place, perhaps through the agency of primitive organisms. As soon as free oxygen appeared in the atmosphere, 1,000,000,000 to 2,000,000,000 years ago, the geochemical cycle for iron was profoundly modified, and this type of transportation and deposition ended forever.

Processes other than fractional crystallization from igneous melts also give rise to magmatic ore deposits. Economic deposits of the oxide mineral chromite ($[Fe,Mg]$ $[Cr,Al]_2O_4$), for example, occur almost entirely as bands or lenses in magnesium-rich igneous rocks. Chromite evidently crystallizes early from magma, and, being of higher density than the liquid, it sinks to the bottom of the magma chamber and becomes concentrated as almost pure bodies of this mineral. Some accessory minerals of igneous rocks are important sources of metallic elements, but the rocks cannot be mined directly because the grade is too low. If these minerals are chemically and mechanically resistant, weathering and transportation may eventually concentrate them into workable deposits. A large proportion of the world's zirconium, hafnium, rare earths, and thorium, and some iron and titanium, come from such deposits in river and beach sands.

A large number of important ore deposits occur in metamorphic rocks. The ultimate origin of these deposits is frequently obscured by the complex processes they have undergone. If it can be established that the enclosing metamorphic rocks were of sedimentary origin, the question then arises whether the ore material was deposited along with the sediments or was introduced by circulating solutions during the metamorphism or possibly at some later time. The answer is seldom clear-cut, and such deposits continue to excite lively controversy among geochemists and economic geologists.

Mineral Fuels

The mineral fuels—coal, petroleum, and natural gas—may be described as a special type of economic deposit. Geochemically they represent the concentration of carbon and hydrogen by processes that were initially biological in nature. Coal is essentially the product of accumulation of land plants in large amounts, and petroleum and natural gas are the products of marine organisms (although the origin of some petroleum and natural gas under non-marine conditions cannot be entirely excluded). The origin of petroleum and natural gas presents a more difficult problem than coal because they are fluids and thus are free to migrate from their place of origin.

The formation of coal is a relatively straightforward geochemical process that can readily be traced through its successive stages. The first requirement is a geological one—the rapid accumulation of plant material under conditions that inhibits its decomposition, followed by its burial under inorganic sediments such as shales and sandstones. The great coal-forming period in the Northern Hemisphere followed the Devonian Period (345,000,000 to 395,000,000 years ago), when abundant land plants first appeared, and has been named the Carboniferous Period (280,000,000 to 345,000,000 years ago). During this period, large areas in North Americaand Europe were evidently low-lying swamps that supported lush vegetation. This vegetation died, accumulated in successive layers, and was partly decomposed by bacteria and other organisms to form peat. Burial of peat deposits under inorganic sediments brought an end to the period of bacterial decomposition, and the further changes to coal were essentially a mild metamorphism caused by an increase in temperature and pressure.

Chemically, this mild metamorphism was in large part the expulsion of carbon dioxide and water from the coal-forming substance. The main trend in the change from peat through lignite to bituminous coal and anthracite is the decrease in oxygen content and the increase in carbon. If carried to its ultimate conclusion the product would be pure carbon in the form of graphite. This occurs comparatively rarely, but evidence for it is the presence of small amounts of graphite in many metamorphic rocks.

Coal also contains inorganic material that appears as ash when it is burned, and some coal ashes show a remarkable concentration of unusual elements. This was demonstrated by Goldschmidt when he found appreciable amounts of germanium in some coal ashes. The Hartley Seam of the Durham Coalfield in England contains so much germanium that the ash has a brilliant yellow colour because of the presence of the oxide (GeO_2).

The source of these minor and trace elements and their mode of incorporation in the coal are still not fully understood. There are three possibilities: (1) these elements were taken up by the plants during growth; (2) they were carried into the coal swamp as a component of the inorganic sediments; or (3) they were absorbed during or after the coal-forming processes from circulating solutions. The first possibility is not favoured, because growing plants seldom incorporate appreciable amounts of nonorganic elements. The second possibility also is unlikely, because there is no correlation, or rather an inverse correlation, between ash content and trace element concentrations. This leaves the third possibility as the most likely one. The presence of a large amount of carbonaceous matter means that the coal-forming environmentis a highly reducing one, which will favour the precipitation of some elements; the presence of hydrogen sulfide and sulfide ions will cause the precipitation of chalcophile elements (with affinity for sulfur); and complex organic compounds are noted for their absorptive, or chelating, capacity for metallic ions. Thus several individual reactions are potentially available for the fixation of foreign elements in the coal substance.

The origin of petroleum is not as readily elucidated as the origin of coal, because petroleum can migrate from the region in which it was formed. Indeed, the very occurrence of a commercial oil field probably implies the concentration of the petroleum from a large volume of source rocks into a relatively much smaller reservoir.

The fact that petroleum is almost always found in marine sedimentary rocks has long been a basic argument in favour of a marine origin for this material. It is certainly true that some oil has been found in igneous and metamorphic rocks, but migration from a sedimentary source bed is a reasonable explanation for these occurrences. Proof of a marine origin has been forthcoming in recent years by the sensitive analyses of recent marine sediments, which show that they contain small amounts of petroleum hydrocarbons, evidently generated either directly by marine organisms or by their subsequent decomposition.

Natural petroleum is a complex mixture of hundreds of different hydrocarbons, but its bulk composition is remarkably constant, about 85 percent carbons and 15 percent hydrogen. It may include small amounts of organic compounds containing oxygen, sulfur, and nitrogen. Its content of other elements is exceedingly small. Petroleum ash, unlike coal ash, is not noted for its trace element content. Some petroleum ash contains appreciable amounts of vanadium, however, and has been utilized as a source of this element. A class of nitrogen-bearing organic compounds known as porphyrins include a metal atom in their molecular structure; usually this atom is iron, but other elements in this region of the periodic table, especially vanadium, nickel, and copper, may play a similar role. The vanadium content of some petroleum ash probably originates as a vanadium porphyrin in some of the organisms involved in petroleum formation.

Petroleum is always accompanied by natural gas, but many natural gas fields have no petroleum associated with them. This can probably be ascribed to greater possibility for migration for a gas as compared to a liquid. It is also possible that some natural gas is generated from coal deposits. Natural gas consists largely of methane, but small amounts of more complex hydrocarbons may be present, and it may also contain unwanted components such as nitrogen, carbon dioxide, and hydrogen sulfide. Natural gas containing hydrogen sulfide is known as "sour gas" and for long was an undesirable material because of the noxious nature of this compound; recently, however, it has been found profitable to extract the hydrogen sulfide by converting it to sulfur and then utilize the hydrocarbons.

Natural gas is the sole source of one element, helium, the industrial demand for which has steadily increased in recent years. Comparatively few occurrences of natural gas contain sufficient helium for the extraction to be commercially profitable. Currently, the Western world's need for helium is largely met by its extraction from wells in western Texas.

Soils

Soil is a thin veneer that forms a discontinuous cover on the land areas of the Earth. Its volume and its mass are small in comparison to the major geospheres, but it is of vast importance to man. Superficially it might be considered merely as comminuted (pulverized) and decomposed bedrock; however, this viewpoint takes into account only its inorganic components and completely neglects the complex of organic compounds, living organisms, water, and included gases that gives the soil its characteristic properties and its value as the abode of life. Comminuted bedrock alone is not soil in any real sense; the Antarctic continent, where not ice-covered, has a surface of comminuted bedrock, as does the Moon, but neither material can properly be termed soil.

Soils result from the weathering of rocks, and hence their composition might be expected to reflect the composition of the rocks from which they were formed. This is true only in a very broad sense, however. Environmental factors play an important part in soil formation. The same parent rock may give rise to very different soils under different conditions. Climate, topography, vegetation, biological activity, and time are all important factors in determining the nature of a soil. Climate is probably the most important of these, as can be demonstrated by contrasting the soils developed on the same rock type under tropical and temperate conditions. In general, the soil in the humid tropics will be different in texture and composition and much less fertile, as a result of the intense leaching brought about by high rainfall, high temperatures, and the almost complete removal of organic matter by microorganisms.

The complex of inorganic compounds, organic compounds, water, and air that makes up the soil is in a continual state of change. Water tends to dissolve and remove the relatively soluble elements such as calcium, magnesium, sodium, and potassium, and the comparatively insoluble elements—aluminum, iron, and silicon—are thereby relatively enriched in the soil. The enrichment of iron is frequently manifested by a red-brown or yellow-brown colour caused by an accumulation of iron oxides. The most reactive part of the soil is the complex of clay minerals and organic matter, which is largely responsible for its agronomic characteristics. True soil does not exist without the presence of colloidal and organic matter. The relative absence of soils in desert areas reflects the fact that chemical and physical weathering of rocks alone does not necessarily result in soil formation. Most soil processes are directly or indirectly biological in nature. Organisms and organic compounds produced by their vital activities or their decomposition are effective agents for dissolving and extracting many elements from the inorganic constituents of the soil, thereby making them available for plant growth.

Although soils do differ in composition, the range of variation in the major elements is rather small. Minor and trace elements may show considerably greater variability. The importance of certain trace elements in the soil for the healthy growth of plants, and through the plants, of the animals that graze on them, has become increasingly apparent in recent years. Most soils contain these trace elements in sufficient amounts, but when deficiencies are present, puzzling diseases appear which in the past have rendered large areas of otherwise suitable land unavailable for farming. On a large area in the North Island of New Zealand, for example, although it grew satisfactory pasture, sheep and cattle failed to thrive and eventually died if not removed. As a result, much of this area was given over to afforestation. It was eventually discovered that cobalt, in the amount of a few parts per million, would completely eliminate the disease when applied in fertilizer or administered directly to the animals. The ultimate explanation is the need of animals (but not plants) for vitamin B_{12}, which contains an atom of cobalt in its structure.

Occasionally, an excess of a specific element may have a deleterious effect on plant growth. Most obvious, of course, are the alkaline or saline soils of desert and coastal areas on which only impoverished vegetation exists. Magnesium-rich soils are notably infertile; such soils develop on areas of ultrabasic igneous rocks consisting largely of olivine, $(Mg,Fe)_2SiO_4$, and the boundaries of these areas can frequently be readily mapped from aerial photographs by the marked change in vegetation. Sometimes plants take up available trace elements in amounts deleterious to animals grazing on them. A well-known example is Astragalus racemosus (locoweed), which in some areas of the western U.S. contains sufficient selenium to be poisonous to grazing animals.

The possible correlation between soil geochemistry and the geographical distribution of disease is thus a field of extreme significance which as yet has been insufficiently studied. The problem is a complex one, in large part because of the difficulty in isolating the numerous factors involved.

The Hydrosphere

The hydrosphere is the discontinuous shell of water—fresh, salt, and solid—on the surface of the Earth. As such it comprises the oceans and the connecting seas and inlets, the lakes, rivers, and streams, the groundwater that feeds them, and the snow and ice cover of high altitudes and high latitudes. The mass of the ocean waters far outweighs the other parts of the hydrosphere. Goldschmidt estimated that there are 273 litres of water in all its forms for every square centimetre of the Earth's surface made up as follows:

	Litres	Kilograms	Total mass
Seawater	268.4	278.1	1.4×10^{18} tons
Freshwater	0.1	0.1	5.1×10^{14} tons
Continental ice	4.5	4.5	2.3×10^{16} tons

Seawater thus makes up over 98 percent of the total mass of the hydrosphere, and its composition essentially can be taken as giving the average composition of the hydrosphere.

Composition of Seawater

Research during the past century has demonstrated that the composition of seawater is essentially uniform and that the relative proportions of the various ions are practically constant. In the open ocean the salinity (approximately the total weight of dissolved solids per kilogram) averages about 35 parts per thousand, but may rise to 40 parts per thousand in regions such as the Red Sea and the Persian Gulf, where rainfall and inflow are low and evaporation high. Sodium chloride is the dominant compound of the salts in solution and comprises about three-quarters of the whole; the remainder consists largely of chlorides and sulfates of magnesium, calcium, and potassium.

Though many data on minor and trace elements in seawater are available, the interpretation of these data is subject to some uncertainties. The concentrations of these elements are probably more variable than for the major elements and may depend to some degree on the sampling location. This is particularly true for the elements that are utilized by marine organisms. Phosphorus is a good example; it is markedly depleted in surface and near-surface waters by biological activity, but it enriches the deeper parts of the ocean through the dissolution of dead organisms. Silica is brought into the ocean in large amounts in solution in river water, but most of it is soon removed to become the skeletal material of diatoms, radiolaria, and sponges.

Seawater also contains dissolved gases in variable amounts. Seawater of normal salinity at 0 °C (32 °F) in equilibrium with the atmosphere will contain about eight millilitres per litre of dissolved oxygen and 14 millilitres per litre of dissolved nitrogen. Dissolved nitrogen is essentially an inert constituent, but dissolved oxygen plays a fundamental role in the growth and decay of organisms and so varies greatly in concentration from place to place. In stagnant regions that are rich in decaying organic matter the water may be completely depleted in free oxygen, and a considerable concentration of hydrogen sulfide may be present. Much of the Black Sea below a depth of a few hundred metres is in this condition.

Another dissolved gas of prime importance for biological activity is carbon dioxide. The conditions controlling its concentration are quite complex, however, because in solution it can be present as free carbon dioxide, as undissociated carbonic acid, as carbonate ions, and as bicarbonate ions. The concentration of these ions will also be affected by biological activity and by the precipitation or dissolution of calcium carbonate.

Circulation of Water Through the Hydrosphere

The circulation of water through the hydrosphere is controlled in large part by the reservoir effect of the oceans. Evaporation from the ocean surface is precipitated as rainfall. Of that falling on the land, some is directly re-evaporated, some is absorbed into the reservoir of groundwater, and some flows off directly into rivers and streams. The total annual rainfall on the Earth is estimated to be 123×10^{18} grams, and the total annual runoff to the oceans 32×10^{18} grams.

Even the purest rainfall contains some material in solution, not only dissolved gases but also nonvolatile material. Rainfall near seacoasts always contains some sodium chloride and small amounts of other marine salts, the concentration of which falls off generally with distance from the ocean. Rainfall in industrial regions may of course contain a variety of pollutants; in many areas it is essentially a dilute sulfuric acid solution. Such material may also be carried far beyond the place of origin; acid rainfall in the Scandinavian countries probably originates in part from England and Germany.

The runoff from the land contains additional material in solution, picked up during its circulation through the crustal rocks. River water averages about 120 parts per million dissolved solids, but the range is great, from about 10 parts per million up to several thousand parts per million. Commonly, the range is from 50 to 200 parts per million; contents greater than 200 parts per million are usually the result of human activities or of drainage from soils containing soluble salts, as in desert regions.

With an average content of 120 parts per million dissolved matter, the rivers of the world deliver 3.9×10^9 tons of material in solution to the sea each year. The average concentration of the important constituents (in parts per million) is: bicarbonate, 58.4; sulfate, 11.2; chloride, 7.8; nitrate, 1.0; calcium, 15.0; magnesium, 4.1; sodium, 6.3; potassium, 2.3; iron, 0.67; and silica, 13.1. Although these ten constituents account for most of the dissolved material, many other elements have been detected in river and lake waters.

Geochemical Balance of Seawater Over Time

The 3.9×10^9 tons carried annually in solution to the oceans are but a small fraction of the total amount of material in solution in the oceans. Nevertheless, when integrated over the whole of geological time, more than 4×10^9 years, it greatly exceeds the present material in solution. Some of the material, especially sodium chloride, is of course cyclical, being circulated from the oceans to the land as aerosols and incorporated in marine sedimentary rocks and ultimately in large part being returned to the oceans in runoff.

Goldschmidt made an interesting calculation on the geochemical balance in seawater. From the amount and composition of sedimentary rocks he estimated that erosion during geological time had amounted to about 160 kilograms of igneous rock per square centimetre of the Earth's surface. Combining this figure with the amount of seawater per square centimetre, 273 kilograms, he derived a figure of 600 grams of igneous rock eroded per kilogram of seawater. Assuming this 600 grams had gone fully into solution (obviously a gross simplification but a limiting one), he drew up a balance sheet between the amounts of different elements potentially supplied to the oceans and the amounts actually present. Some of these figures are presented in the table. Despite the imperfections of the method, the results are certainly significant in a qualitative sense. Some elements—chlorine, bromine, boron, and sulfur—are present in seawater in amounts far in excess of those that can have been derived by erosion. The source of these "superabundant" elements has probably been volcanism and related magmatic activity. Halides, sulfates, and borates are deposited by volcanic gases and carried in solution in hot springs. The relative depletion of fluorine with respect to chlorine in seawater can be ascribed to the precipitation of highly insoluble fluorine-bearing compounds, mainly apatite (calcium fluophosphate). Sodium clearly remains in solution to a much greater extent than potassium; the latter element reacts with sedimentary materials to form insoluble potassium-bearing silicates such as illite and glauconite, which have no sodium-bearing analogs. Calcium is removed from solution much more effectively than strontium, evidently because it is utilized by organisms. Goldschmidt pointed out that many highly poisonous elements, such as arsenic and selenium, have been potentially supplied in dangerous amounts. Their concentration remains very low, however, presumably because of efficient processes of removal as insoluble compounds. Adsorption on colloidal particles of clay and iron oxides is a likely process.

Geochemical balance of some elements in seawater			
Element	Potential amount supplied to oceans (g/ton)	Amount present in seawater (g/ton)	Percentage in solution
Lithium	39	0.17	0.4
Boron	2	4.5	250
Fluorine	540	1.3	0.2
Sodium	16,980	10,800	64
Magnesium	12,540	1,290	10
Phosphorus	708	0.09	0.01
Sulfur	312	904	290
Chlorine	188	19,400	10,300
Potassium	15,540	392	2.5
Calcium	21,780	411	1.9
Arsenic	3	0.003	0.1
Bromine	0.97	67	6,900
Rubidium	186	0.12	0.06
Strontium	180	8.1	4.6
Iodine	0.18	0.06	33
Cesium	4	0.0003	0.008
Barium	150	0.02	0.01

The geological and geochemical evidence indicates that the ocean waters are, and have been for a long time, in a steady state of essentially unchanging composition. The addition of material by runoff from the land is adjusted by reactions within the ocean waters or between the ocean waters and sedimentary materials whereby the concentrations of the individual elements remain essentially constant. How far back in geological time this steady state has persisted remains something of an open question. The existence of most forms of marine life from the Cambrian to the present indicates a uniformity of marine conditions over the past 600,000,000 years; how far back into the Precambrian this uniformity extended is more difficult to elucidate.

The Atmosphere

The atmosphere is the most homogeneous and thus the most easily studied of the geospheres. Its mass is readily determined from the product of the average height of the mercury barometer in centimetres, the density of mercury (13.6 grams per cubic centimetre), and the area of the Earth (5.1×10^{18} square centimetres). Recent calculations give 51.17×10^{20} grams for its total mass.

Composition

The composition is also relatively simple, although a considerable number of gases may be present in small amounts. Almost 99 percent consists of oxygen and nitrogen, with argon making up most of the remainder. Carbon dioxide, essential for plant life, is present in an extremely small amount. Some gases not listed in the table may be present as local or even regional pollutants—city dwellers are becoming increasingly aware of oxides of sulfur as atmospheric pollutants, and the scientific study of smog is largely concerned with reactions taking place between hydrocarbons, oxides of nitrogen, oxygen and ozone.

Average composition of the atmosphere			
Gas	Composition by volume (ppm)*	Composition by weight (ppm)*	Total mass (10^{20} g)
Nitrogen	780,900	755,100	38.648
Oxygen	209,500	231,500	11.841
Argon	9,300	12,800	0.655
Carbon dioxide	386	591	0.0299
Neon	18	12.5	0.000636
Helium	5.2	0.72	0.000037
Methane	1.5	0.94	0.000043
Krypton	1.0	2.9	0.000146
Nitrous oxide	0.5	0.8	0.000040
Hydrogen	0.5	0.035	0.000002
Ozone**	0.4	0.7	0.000035
Xenon	0.08	0.36	0.000018

*ppm = parts per million.
**Variable, increases with height.

The atmosphere gradually thins out into the vacuum of outer space, and its upper limit can conveniently be placed at about 600 kilometres. An important zone in the stratosphere is known as the ozonosphere, a diffuse layer characterized by an increase in the concentration of ozone, O_3. This zone is highly important for life on Earth because it absorbs most of the ultraviolet radiation from the Sun; if this penetrated to the Earth's surface it would act as a potent sterilizer, fatal for most forms of life. It also helps to maintain a more uniform surface temperature by reducing the loss of heat by radiation to space—the so-called greenhouse effect.

Geochemical History

The geochemical history of the atmosphere has been a complex one. Scientists agree that the present atmosphere is quite different from the original one. It is certainly quite different from those of the other planets. It is reasonable to conclude that this reflects, in part at least, the Earth as the abode of life. The Earth's atmosphere differs from those of its neighbours in the solar system probably in large part through the action of photosynthesis, a complex biological process which was probably preceded by a lengthy period of organic evolution.

The nature of the Earth's primitive atmosphere is still a subject of some speculation. Some scientists, reasoning by analogy with the larger planets such as Jupiter, have argued for an original atmosphere consisting largely of methane and ammonia. Others have considered that present-day volcanic gases may indicate the nature of the primitive atmosphere, in which case it contained carbon dioxide, possibly carbon monoxide, nitrogen, and water vapour. In either case, free oxygen was absent. If the evolution of the atmosphere is traced backward in time through the geological record, then extensive terrestrial photosynthesis is indicated by an abundance of land plants in Devonian times, about 400,000,000 years ago. Marine photosynthesis, however, is much older, since practically all the major groups of marine organisms were established by the beginning of the Cambrian period, some 540,000,000 years ago. The extensive Precambrian iron formations suggest an oxygen-free atmosphere which was terminated about 2,000,000,000 years ago. This evidence has been translated into estimates of oxygen content of the atmosphere of about 1 percent of the present level 2,000,000,000 years ago, about 10 percent of the present level at the beginning of the Cambrian Period, and essentially the present content by Devonian times.

Although it is not yet possible to know the quantitative composition of the primitive atmosphere, the geochemical processes that have operated to modify its composition during geological time can be evaluated. These processes can be summarized as a series of gains and losses. Additions to the atmosphere comprise: (1) gases released by igneous activity; (2) oxygen and hydrogen produced by the photochemical dissociation of water vapour; (3) oxygen produced by photosynthesis; (4) helium produced by the radioactive breakdown of uraniumand thorium; and (5) argon produced by the radioactive breakdown of potassium. Atmospheric losses include: (1) oxygen removal by oxidation of ferrous to ferric iron, sulfur compounds to sulfates, hydrogen to water, and similar reactions; (2) carbon dioxide removed by the formation of coal, petroleum, and the death and burial of organisms; (3) carbon dioxide removed by the formation of calcium and magnesium carbonates; (4) nitrogen removed by the formation of oxides of nitrogen in the air and by the action of nitrifying bacteria in the soil; and (5) hydrogen and helium by escape from the Earth's gravitational field.

Photosynthesis has certainly been the most significant process in controlling atmospheric composition during much of geological time. Through this process carbon dioxide and water are converted to carbohydrate, with the accompanying release of oxygen. Much of this carbohydrate is consumed by animals and reconverted to carbon dioxide and water by respiration, and oxidative decay leads to the same result. Some, however, is incorporated into sediments; part may go to form exploitable deposits of coal and petroleum, but most of it remains as disseminated carbonaceous material; the average carbon content of sedimentary rocks is about 0.4 percent.

Quantitatively, more significant amounts of carbon dioxide have been removed from the atmosphere in the form of limestone and dolomite. Most of this removal has been effected by marine organisms, especially algae and corals, but direct inorganic precipitation may occur, especially in warm tropical waters. Judging from the vast deposits of limestone and dolomite throughout the sedimentary record, this process has operated with a remarkable degree of uniformity throughout geological time. Extensive limestone formations are possibly less common in older Precambrian rocks, indicating a slower beginning of carbonate precipitation. It is truly remarkable that so much carbonate rock has been deposited during geological time, with the carbonate being ultimately derived from an atmosphere which may never have contained a much higher concentration of carbon dioxide than is present today. It has been pointed out that reactions like the decomposition of calcium silicate—$CaSiO_3 + CO_2 = CaCO_3 + SiO_2$, in which $CaSiO_3$ is calcium silicate, CO_2 is carbon dioxide, $CaCO_3$ is calcium carbonate, and SiO_2 is silicon dioxide, which tends to go toward the right at ordinary temperatures—will act as buffering mechanisms to keep the carbon dioxide concentration of the atmosphere at a continuously low level.

If the carbon dioxide concentration has remained essentially constant, and yet this compound has been continuously extracted to form carbonates and organic compounds, then clearly a balancing source of "new" carbon dioxide is required. This evidently has been provided by volcanism and other igneous activity. The Earth is steadily being degassed, in the sense that gaseous compounds contained in the mantle are escaping to the surface. The presence of carbon dioxide in the mantle has been demonstrated by the presence of microscopic inclusions of liquid carbon dioxide in the minerals of the peridotite xenoliths (rocks contained within other rocks) brought up in some volcanoes. Along with carbon dioxide, much water and small amounts of other volatiles are being added continuously from sources in the mantle. Ultimately, the hydrosphere, as well as the atmosphere, is the product of the degassing of the Earth's interior.

Of the remaining atmospheric gases, argon presents some intriguing features. Argon is by far the most abundant of the inert gases on Earth, whereas in the universe as a whole it is much less abundant than either helium or neon. In addition, its isotopic composition is quite distinct, consisting almost entirely of argon-40, whereas in the Cosmos argon-36 is the most abundant isotope. The reason for these anomalies is that atmospheric argon is almost entirely radiogenic, the product of the decay of the potassium-40 isotope of potassium.

Similarly, the helium in the atmosphere is probably entirely the product of the radioactive decay of uranium and thorium. Actually, the atmosphere contains only about 10 percent of the total amount of helium generated from these sources during geological time. Some of this helium remains occluded in the rocks where it was formed; some has escaped from the upper atmosphere. Helium (and hydrogen), consisting of light atoms, can escape from the gravitational field of the

Earth, whereas heavier gases cannot. A minor source of atmospheric oxygen throughout geological time is probably the photochemical decomposition of water vapour in the upper atmosphere, with the subsequent loss of the hydrogen to outer space.

Some oxygen has been removed from the atmosphere by oxidative reactions, of which the most significant has been the conversion of ferrous to ferric iron. In igneous rocks the average ferrous-to-ferric iron ratio (FeO/Fe_2O_3) is greater than unity, whereas in sedimentary rocks the proportion is reversed, ferric iron being dominant over ferrous iron. Other oxidative reactions are the conversion of manganous compounds to manganese dioxide and of hydrogen sulfideto free sulfur and sulfate. Nitrogen is almost inert geochemically, but a little is fixed as oxides of nitrogen by lightning, and somewhat more by the action of nitrifying bacteria in the soil. Most of this nitrogen is ultimately returned to the atmosphere by the decay of the organisms. Oxides of nitrogen formed in the atmosphere are removed in rain as nitrite and nitrate. Nitrogen does not accumulate in the soil, however, except perhaps under extremely arid conditions, as in the deserts of northern Chile, the locale of the unique nitrate deposits.

Geochemical Cycle

Geochemical Cycle is the developmental path followed by individual elements or groups of elements in the crustal and subcrustal zones of the Earth and on its surface. The concept of a geochemical cycle encompasses geochemical differentiation (*i.e.*, the natural separation and concentration of elements by Earth processes) and heat-assisted, elemental recombination processes.

For the lithosphere (*i.e.*, the crust and upper mantle), the geochemical cycle begins with the crystallization of a magma at the surface or at depth. In turn, surface alteration and weathering break down the igneous rock, a process that is followed by the transportation and depositionof the resulting material as sediment. This sediment becomes lithified and eventually metamorphosed until melting occurs and new magma is generated. This ideal cycle can be interrupted at any point. Each element may be affected differently as the cycle progresses. During the weathering of an igneous rock, for example, minerals containing iron, magnesium, and calcium break down and are carried in solution, but silicon-rich quartz and feldspar are mainly transported as sediment. The resultant sedimentary rocks are dominated by quartz and feldspar, whereas others are dominated by calcium and magnesium owing to the precipitation of calcium or magnesium carbonates. Such elements as sodium remain in solution until precipitated under extreme conditions. As partial melting of sedimentary rocks begins, elements become separated according to melting properties; volatiles are released to the atmosphere, and physical movement of chemically separated bodies occurs. While the geochemical cycle over a short term is in a seemingly steady state, long-term, or secular, changes occur. Thus, for example, continents and oceans have evolved over geologic time.

The Geological Carbon Cycle

In the geological carbon cycle, carbon moves between rocks and minerals, seawater, and the atmosphere. Carbon dioxide in the atmosphere reacts with some minerals to form the mineral calcium carbonate (limestone). This mineral is then dissolved by rainwater and carried to the oceans. Once there, it can precipitate out of the ocean water, forming layers of sediment on the sea floor. As the Earth's plates move, through the processes of plate tectonics, these sediments are subducted underneath the continents. Under the great heat and pressure far below the Earth's surface, the limestone melts and reacts with other minerals, releasing carbon dioxide. The carbon dioxide is then re-emitted into the atmosphere through volcanic eruptions.

Concern of rising emissions: What would happen if the carbon dioxide in Earth's atmosphere increased? The Earth's surface temperatures would increase. If temperatures increased too much, humans (and other forms of life) might not be able to survive.

Cars that pollute the environment less than standard gasoline-powered cars.

At its recent meeting in Bangkok - with international economists, scientists and government officials from more than 100 countries - the United Nations' Intergovernmental Panel on Climate Change (IPCC) raised the concern of rising emissions. The rise in global income and the world's population have added to the burning of oil, coal and gas. As a result, emissions of greenhouse gases have shot up 70% between 1970 and 2004. For carbon dioxide alone, the increase was 80% in the same time period.

The options to tackling emissions are:

- Pricing carbon: Consumers and producers should pay a price for goods and services which reflect the environmental damage from using fossil fuels. This promotes energy efficiency and the switch to cleaner sources.

- Renewable energies: Wind, solar and geothermal power should be promoted through subsidies, preferential tariffs and rules requiring buying from these sources.

- Energy efficiency: Tougher building standards, mandatory fuel economy, biofuel blending and investment in better public transport.

- Carbon sequestration: Building underground chambers to store CO_2 emissions from coal-fired plants.

- Nuclear power: Wider use of nuclear energy, up from 16% of electricity supply in 2005 to 18% share by 2030.

The Geological Carbon Cycle

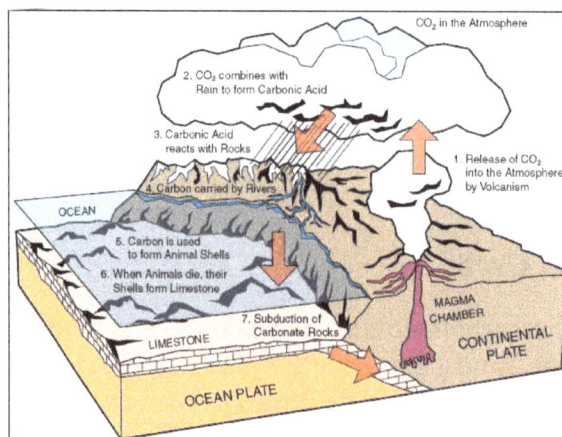

Geophysics

Geophysics is the major branch of the Earth sciences that applies the principles and methods of physics to the study of the Earth.

Geophysics deals with a wide array of geologic phenomena, including the temperature distribution of the Earth's interior; the source, configuration, and variations of the geomagnetic field; and the large-scale features of the terrestrial crust, such as rifts, continental sutures, and mid-oceanic ridges. Modern geophysical research extends to phenomena of the outer parts of the Earth's atmosphere (e.g., the ionospheric dynamo, auroral electrojets, and magnetopause current system) and even to the physical properties of other planets and their satellites.

Many of the problems of geophysics are analogous to those of astronomy because the subject studied is rarely under direct observation, and conclusions must be drawn largely on the basis of mathematical interpretation of physical measurements. These include measurements of the Earth's gravitational field with gravimeters on land and sea and artificial satellites in space; magnetometric measurements of the planet's magnetic field; and seismological surveys of subsurface geologic structures using reflected and refracted elastic waves produced by earthquakes or by artificial means.

Research conducted with geophysical techniques has proved extremely useful in providing evidence in support of the theory of plate tectonics. Seismographic data, for instance, have demonstrated that the world's earthquake belts mark the boundaries of the enormous rigid plates that constitute the Earth's outer shell, while the findings of paleomagnetic studies have made it possible to trace the drift of the continents over geologic time.

Earth's Shape and Size

Earth's Shape

Earth's circumference and diameter differ because its shape is classified as an oblate spheroid or ellipsoid, instead of a true sphere. This means that instead of being of equal circumference in all areas, the poles are squished, resulting in a bulge at the equator, and thus a larger circumference and diameter there.

The equatorial bulge at Earth's equator is measured at 26.5 miles (42.72 km) and is caused by the planet's rotation and gravity. Gravity itself causes planets and other celestial bodies to contract and form a sphere. This is because it pulls all the mass of an object as close to the center of gravity (the Earth's core in this case) as possible.

Because Earth rotates, this sphere is distorted by the centrifugal force. This is the force that causes objects to move outward away from the center of gravity. Therefore, as the Earth rotates, centrifugal force is greatest at the equator so it causes a slight outward bulge there, giving that region a larger circumference and diameter.

Local topography also plays a role in the Earth's shape, but on a global scale, its role is very small. The largest differences in local topography across the globe are Mount Everest, the highest point above sea level at 29,035 ft (8,850 m), and the Mariana Trench, the lowest point below sea level at 35,840 ft (10,924 m). This difference is only a matter of about 12 miles (19 km), which is quite minor overall. If the equatorial bulge is considered, the world's highest point and the place that is farthest from the Earth's center is the peak of the volcano Chimborazo in Ecuador as it is the highest peak that is nearest the equator. Its elevation is 20,561 ft (6,267 m).

Earth's Size

As the largest of the terrestrial planets, Earth has an estimated mass of 5.9736×10^{24} kg. Its volume is also the largest of these planets at 108.321×10^{10} km³.

In addition, Earth is the densest of the terrestrial planets as it is made up of a crust, mantle, and core. The Earth's crust is the thinnest of these layers while the mantle comprises 84% of Earth's volume and extends 1,800 miles (2,900 km) below the surface. What makes Earth the densest of these planets; however, is its core. It is the only terrestrial planet with a liquid outer core that surrounds a solid, dense inner core. Earth's average density is 5515×10 kg/m³. Mars, the smallest of the terrestrial planets by density, is only around 70% as dense as Earth.

Earth is classified as the largest of the terrestrial planets based on its circumference and diameter as well. At the equator, Earth's circumference is 24,901.55 miles (40,075.16 km). It is slightly smaller between the North and South poles at 24,859.82 miles (40,008 km). Earth's diameter at the poles is 7,899.80 miles (12,713.5 km) while it is 7,926.28 miles (12,756.1 km) at the equator. For comparison, the largest planet in Earth's solar system, Jupiter, has a diameter of 88,846 miles (142,984 km).

Geodesy

Geodesy is the science which deals with the methods of precise measurements of elements of the surface of the earth and their treatment for the determination of geographic positions on the surface of the earth. It also deals with the theory of size and shape of the earth. Geodesy may be broadly divided into two branches, namely:

1. Geometric Geodesy

2. Physical Geodesy

There is third one also, known as Satellite Geodesy. Dictionary meaning of Geodesy is 'dividing the earth and measurement of the earth geometry'. Thus geometric geodesy appears to be purely geometrical science as it deals with the geometry (shape and size) of the earth.

Determination of geographical positions on the surface of earth can be made by observing celestial bodies, and thus comes under geodetic astronomy, but this can be included under geometric geodesy.

Earth gravity field is a physical entity and is involved in most of the geodetic measurements, even the purely geometric ones. The measurement of geodetic astronomy, triangulation and leveling, all make essential use of plumb line being the direction of gravity vector.

Thus, astro-geodetic methods which use astro determination of latitude, longitude, and azimuth and geodetic operations e.g. triangulation, trilateration, base measurement etc., may be considered as belonging to Physical geodesy fully as much as the gravimetric methods.

As a general distinction astro-geodetic methods come under geometric geodesy which use the direction of gravity vector, employing geometrical techniques, whereas the gravimetric methods come under physical geodesy, which operate with the magnitude of 'g' using potential theory. A sharp demarcation is impossible and there are frequent overlaps.

Satellite Geodesy: Satellite geodesy comprises the observational and computational techniques which allow the solution of geodetic problems by the use of precise measurements to, from or between artificial, mostly near the earth satellites.

Applications of Geodesy

1. Primary or Zero order triangulation, trilateration and traverse.

2. The measurement of height above sea-level by triangulation or sprit leveling.

3. Astronomically observations of latitude, longitude and azimuth to locate origins of surveys, and to control their direction.

4. Crustal Movements. To detect changes in the relative positions on the ground, and in their heights above sea level.

5. Observation of the direction of Gravity by astronomical observations for latitude and longitude.

6. Observation of the intensity of Gravity by the pendulum and other apparatus.

7. To deduce the exact form of earth's sea level equipotential surfaces at all heights.

8. Polar motion studies.

9. Earth tides.

10. The separation between the Geoid and the mean sea level.

11. Engineering Surveys.

12. Satellite Geodesy: includes the modern techniques of positioning by space methods e.g. GPS, SLR and VLBI etc.

The Importance Reference Surfaces in Geodesy

To understand how the shape and size of the earth is determined, three surfaces which are of interest to the geodesists have to be understood clearly. These surfaces are:

1. Physical surface of the earth,

2. Geoid,

3. Reference Ellipsoid.

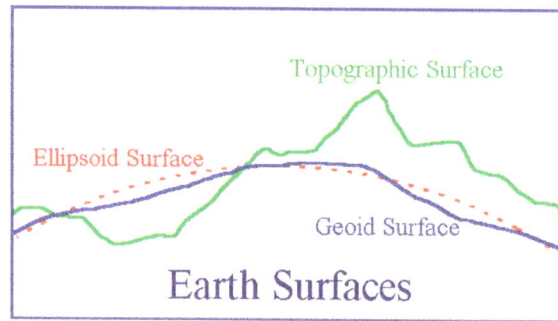

Three Surfaces of Earth

Geomagnetic Field

The Earth's magnetic field is similar to that of a bar magnet tilted 11 degrees from the spin axis of the Earth. The problem with that picture is that the Curie temperature of iron is about 770°C. The Earth's core is hotter than that and therefore not magnetic. So how did the Earth get its magnetic field.

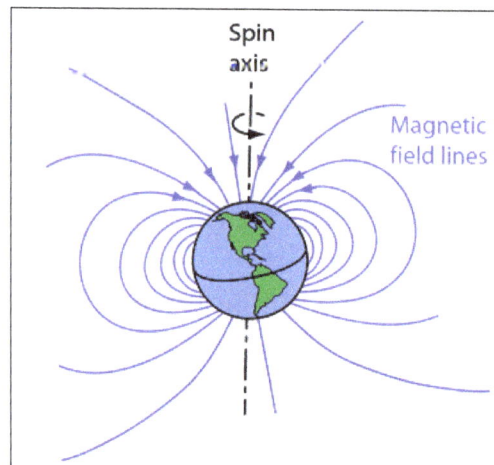

Magnetic fields surround electric currents, so we surmise that circulating electic currents in the Earth's molten metalic core are the origin of the magnetic field. A current loop gives a field similar to that of the earth. The magnetic field magnitude measured at the surface of the Earth is about half a Gauss and dips toward the Earth in the northern hemisphere. The magnitude varies over the surface of the Earth in the range 0.3 to 0.6 Gauss.

The Earth's magnetic field is attributed to a dynamo effect of circulating electric current, but it is not constant in direction. Rock specimens of different age in similar locations have different directions of permanent magnetization. Evidence for 171 magnetic field reversals during the past 71 million years has been reported.

Although the details of the dynamo effect are not known in detail, the rotation of the Earth plays a part in generating the currents which are presumed to be the source of the magnetic field. Mariner 2 found that Venus does not have such a magnetic field although its core iron content must be similar to that of the Earth. Venus's rotation period of 243 Earth days is just too slow to produce the dynamo effect.

Interaction of the terrestrial magnetic field with particles from the solar wind sets up the conditions for the aurora phenomena near the poles.

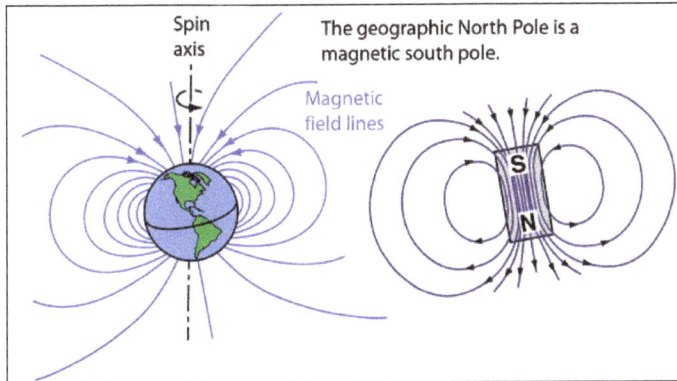

The north pole of a compass needle is a magnetic north pole. It is attracted to the geographic North Pole, which is a magnetic south pole (opposite magnetic poles attract).

The Dynamo Effect

The simple question "how does the Earth get its magnetic field?" does not have a simple answer. It does seem clear that the generation of the magnetic field is linked to the rotation of the earth, since Venus with a similar iron-core composition but a 243 Earth-day rotation period does not have a measurable magnetic field. It certainly seems plausible that it depends upon the rotation of the fluid metallic iron which makes up a large portion of the interior, and the rotating conductor model leads to the term "dynamo effect" or "geodynamo", evoking the image of an electric generator.

Convection drives the outer-core fluid and it circulates relative to the earth. This means the electrically conducting material moves relative to the earth's magnetic field. If it can obtain a charge by some interaction like friction between layers, an effective current loop could be produced. The magnetic field of a current loop could sustain the magnetic dipole type magnetic field of the earth.

Rock Magnetism

Magnetic fields are produced by convection in the Earth's core, and by magnetized rocks. Important geophysical and geological observations can be made from both types of fields.

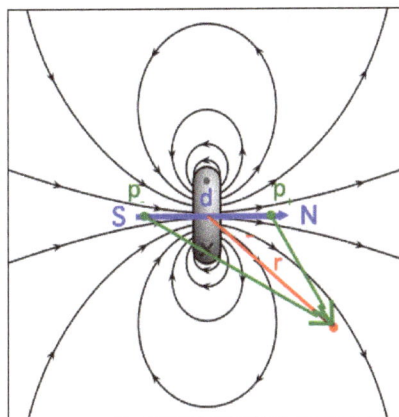

All magnetic fields are produced by electric currents, and are oriented perpendicular to the current. For a loop of current, the magnetic field is that of a magnetic dipole, which is the field produced by two poles of opposite sign (p+ and p-, with units of A m) separated by a small distance d (note that a magnetic monopole is not possible). The force between poles is given by Coulomb's law:

$$F(r) = \frac{\mu_0}{4\pi} \frac{p+p-}{r_2}$$

where $\mu_0 = 4\pi \times 10^{-7}\,N/A^2$ is the permeability constant and r is distance. The force exerted on a unit pole is thus:

$$B(r) = \frac{\mu_0}{4\pi} \frac{p}{r^2}$$

where B is the magnetic field, with units of Tesla (N A^{-1} m^{-1}). The magnetic potential is:

$$W = -\int_r^\infty B dr = \frac{\mu_0}{4\pi} \frac{p}{r}$$

The distances to the p_+ and p_- poles are:

$$r_+ = r - \frac{d}{2}\cos\theta \text{ and } r_- = r + \frac{d}{2}\cos\theta.$$

Simplifying and assuming that $d<<r$, the potential of a magnetic dipole is:

$$W = \frac{\mu_0 p}{4\pi}\left(\frac{1}{r_+} - \frac{1}{r_-}\right) = \frac{\mu_0 p}{4\pi}\left(\frac{r_- - r_+}{r_+ r_-}\right) \approx \frac{\mu_0}{4\pi} \frac{m\cos\theta}{r^2}$$

where m is magnetic moment: m=dp (bar magnet) & m=IA (current loop) Here d is pole separation, p is pole strength, I is current, and A is loop area. In a magnetic field B, the torque on the dipole is:

$$\vec{\tau} = \vec{m} \times \vec{B}$$

Thus, the magnetic moment will orient in the direction of the field (if it is free to do so).

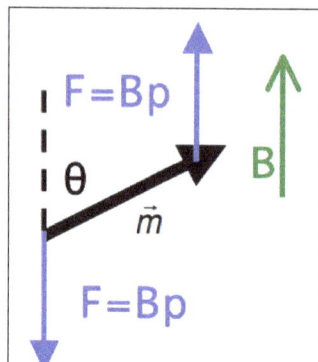

The magnetization $\vec{M}$ of a material is the magnetic moment per unit volume:

$$\vec{M} = \sum \vec{m}_i / V$$

Where, V is the volume

The magnetic field $\vec{B}$ inside of a magnetic material depends on the magnetization of the material. This difference is defined as $\vec{H}$ and is given by:

$$\vec{H} = \vec{B} / \mu_0 - \vec{M}$$

Where, $\vec{H}$ is known as the magnetizing field (A/m).

In a vacuum, $\vec{M} = 0$, so $\vec{B} = \mu_0 \vec{H}$. Inside of a material, the magnetization $\vec{M}$ will affect the net $\vec{B}$ field. However, in general the magnetization is proportional to the magnetizing field:

$$\vec{M} = k\vec{H}$$

Where, k is the magnetic susceptibility. Then:

$$\vec{B} = (1+k)\mu_0\vec{H} = \mu\,\mu_0\,\vec{H}$$

Where, $\mu = (1+k)$ is the magnetic permeability.

The magnetic properties of a material depend on the material's crystal structure and on the presence of unpaired electron spins (that form a net current loop).

1. Diamagnetism: An applied magnetic field exerts a force on orbiting electrons, changing their rotation. This produces a weak induced field that opposes the applied field. Magnetic susceptibility is small and negative $(k \sim -10^{-6}$ for quartz$)$.

2. Paramagnetism (A): An applied field preferentially orients atoms with net electron spins, producing weak magnetization in the direction of the applied field. This tendency is opposed by random atom motions at higher temps. The Curies-Weiss law $k = \dfrac{C}{T-\theta}$ gives susceptibility k where T is temperature, C depends on the material, and θ is the Weiss temperature below which no paramagnetism occurs. $\left(k \sim 10^{-5} \text{ to } 10^{-4} \text{ for olivine, pyroxene at room temp.}\right)$.

3. Ferromagnetism (B): In some metals (iron), electron exchange with neighboring atoms causes magnetic moments to line up, producing a large spontaneous magnetization that disappears above the Curie temperature (TC < θ).

4. Antiferromagnetism (C): In some materials magnetic moments become paired antiparallel, producing no remanent and weak spontaneous magnetization, below the Néel temperature T_N $(T > T_N$ becomes paramagnetic). (ilmenite).

5. Ferrimagnetism (D): Unequal magnetization of lattices and sublattices produces a net spontaneous magnetization and weak remanant magnetization (magnetite).

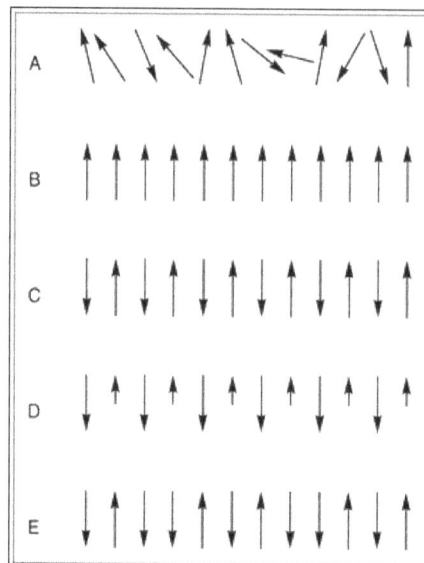

6. Parasitic Ferromagnetism (E): Some iron-rich minerals contain defects or misaligned antiparallel moments that yield a net magnetization (hematite).

7. Remanent Magnetization (Remanence): A permanent magnetization that a ferroor ferri-magnetic material retains after an applied field is removed.

Geologic Magnetizations

1. Thermoremanent Magnetization: A cooling igneous rock retains its spontaneous anisotropy below a "blocking" temperature. Very stable over geologic time.

2. Sedimentary remanent magnetization: Settling ferromagnetic particles align with a magnetic field as they fall, producing a statistical magnetization.

Chemical remanent magnetization can occur when magnetic minerals are altered chemically or precipitate.

References

- Geoscience: environmentalscience.org, Retrieved 2 February, 2019

- CarbonCycleBackground: globecarboncycle.unh.edu, Retrieved 12 April, 2019

- The-sulfur-cycle, biology: lumenlearning.com, Retrieved 20 June, 2019

- Oxygen-cycle-environment, biology: byjus.com, Retrieved 22 April, 2019

- Geochemistry, entry: newworldencyclopedia.org, Retrieved 25 August, 2019

- Geochemical-cycle, science: britannica.com, Retrieved 3 March, 2019

- Geophysics, science: britannica.com, Retrieved 13 July, 2019

- Geodesy-size-shape-of-planet-earth-1435325: thoughtco.com, Retrieved 18 January, 2019

- What-is-geodesy: gisresources.com, Retrieved 1 March, 2019

- MagEarth, magnetic, hbase: phy-astr.gsu.edu, Retrieved 19 May, 2019

Chapter 3

Plate Tectonics

The scientific theory which describes the motion of the different plates of Earth's lithosphere is called plate tectonics. Some of the areas studied within plate tectonics are tectonic landforms, supercontinents, earthquakes and volcanoes. The diverse aspects of plate tectonics as well as these focus areas have been thoroughly discussed in this chapter.

Plate tectonics describes the large scale motions of Earth's lithosphere.

The outermost part of the Earth's interior is made up of two layers: above is the lithosphere, comprising the crust and the rigid uppermost part of the mantle. Below the lithosphere lies the asthenosphere. Although solid, the asthenosphere has relatively low viscosity and shear strength and can flow like a liquid on geological time scales. The deeper mantle below the asthenosphere is more rigid again due to the higher pressure.

The lithosphere is broken up into what are called *tectonic plates* —in the case of Earth, there are seven major and many minor plates. The lithospheric plates ride on the asthenosphere. These plates move in relation to one another at one of three types of plate boundaries: convergent or collision boundaries, divergent or spreading boundaries, and transform boundaries. Earthquakes, volcanic activity, mountain-building, and oceanic trench formation occur along plate boundaries. The lateral movement of the plates is typically at speeds of 50—100 mm/a.

Principles of Plate Tectonics

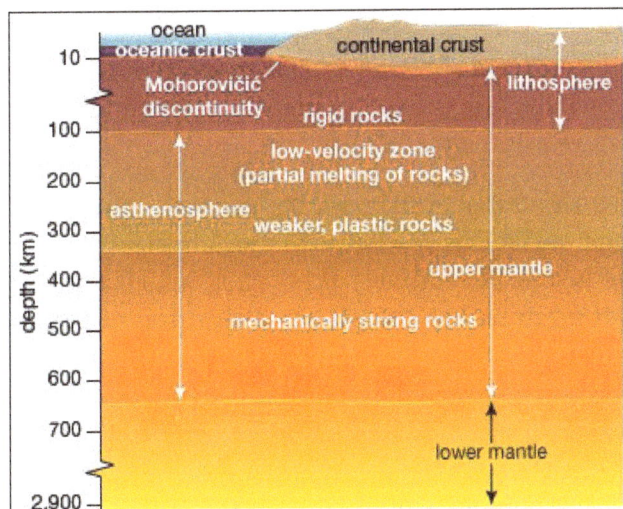

A cross section of Earth's outer layers, from the crust through the lower mantle.

In essence, plate-tectonic theory is elegantly simple. Earth's surface layer, 50 to 100 km (30 to 60 miles) thick, is rigid and is composed of a set of large and small plates. Together, these plates constitute the lithosphere, The lithosphere rests on and slides over an underlying partially molten (and thus weaker but generally denser) layer of plastic partially molten rock known as the asthenosphere, Plate movement is possible because the lithosphere-asthenosphere boundary is a zone of detachment. As the lithospheric plates move across Earth's surface, driven by forces as yet not fully understood, they interact along their boundaries, diverging, converging, or slipping past each other. While the interiors of the plates are presumed to remain essentially undeformed, plate boundaries are the sites of many of the principal processes that shape the terrestrial surface, including earthquakes, volcanism, and orogeny(that is, formation of mountain ranges).

The process of plate tectonics may be driven by convection in Earth's mantle, the pull of heavy old pieces of crust into the mantle, or some combination of both.

Earth's Layers

Knowledge of Earth's interior is derived primarily from analysis of the seismic waves that propagate through Earth as a result of earthquakes. Depending on the material they travel through, the waves may speed up, slow down, bend, or even stop if they cannot penetrate the material they encounter.

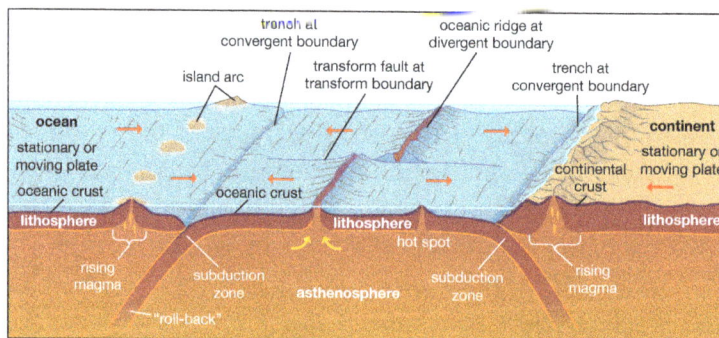

Crustal generation and destruction.

In figure, three-dimensional diagram showing crustal generation and destruction according to the theory of plate tectonics; included are the three kinds of plate boundaries—divergent, convergent (or collision), and strike-slip (or transform).

Collectively, these studies show that Earth can be internally divided into layers on the basis of either gradual or abrupt variations in chemical and physical properties. Chemically, Earth can be divided into three layers. A relatively thin crust, which typically varies from a few kilometres to 40 km (about 25 miles) in thickness, sits on top of the mantle. (In some places, Earth's crust may be up to 70 km [40 miles] thick.) The mantle is much thicker than the crust; it contains 83 percent of Earth's volume and continues to a depth of 2,900 km (1,800 miles). Beneath the mantle is the core, which extends to the centre of Earth, some 6,370 km (nearly 4,000 miles) below the surface. Geologists maintain that the core is made up primarily of metallic iron accompanied by smaller amounts of nickel, cobalt, and lighter elements, such as carbon and sulfur.

There are two types of crust, continental and oceanic, which differ in their composition and thickness. The distribution of these crustal types broadly coincides with the division into continents and ocean basins, although continental shelves, which are submerged, are underlain by continental crust.

The continents have a crust that is broadly granitic in composition and, with a density of about 2.7 grams per cubic cm (0.098 pound per cubic inch), is somewhat lighter than oceanic crust, which is basaltic (i.e., richer in iron and magnesiumthan granite) in composition and has a density of about 2.9 to 3 grams per cubic cm (0.1 to 0.11 pound per cubic inch). Continental crust is typically 40 km (25 miles) thick, while oceanic crustis much thinner, averaging about 6 km (4 miles) in thickness. These crustal rocks both sit on top of the mantle, which is ultramafic in composition (i.e., very rich in magnesium and iron-bearing silicate minerals). The boundary between the crust (continental or oceanic) and the underlying mantle is known as the Mohorovičić discontinuity (also called Moho), which is named for its discoverer, Croatian seismologist Andrija Mohorovičić. The Moho is clearly defined by seismic studies, which detect acceleration in seismic waves as they pass from the crust into the denser mantle. The boundary between the mantle and the core is also clearly defined by seismic studies, which suggest that the outer part of the core is a liquid.

The effect of the different densities of lithospheric rock can be seen in the different average elevations of continental and oceanic crust. The less-dense continental crust has greater buoyancy, causing it to float much higher in the mantle. Its average elevation above sea level is 840 metres (2,750 feet), while the average depth of oceanic crust is 3,790 metres (12,400 feet). This density difference creates two principal levels of Earth's surface.

The lithosphere itself includes all the crust as well as the upper part of the mantle (i.e., the region directly beneath the Moho), which is also rigid. However, as temperatures increase with depth, the heat causes mantle rocks to lose their rigidity. This process begins at about 100 km (60 miles) below the surface. This change occurs within the mantle and defines the base of the lithosphere and the top of the asthenosphere. This upper portion of the mantle, which is known as the lithospheric mantle, has an average density of about 3.3 grams per cubic cm (0.12 pound per cubic inch). The asthenosphere, which sits directly below the lithospheric mantle, is thought to be slightly denser at 3.4–4.4 grams per cubic cm (0.12–0.16 pound per cubic inch).

In contrast, the rocks in the asthenosphere are weaker, because they are close to their melting temperatures. As a result, seismic waves slow as they enter the asthenosphere. With increasing depth, however, the greater pressure from the weight of the rocks above causes the mantle to become gradually stronger, and seismic waves increase in velocity, a defining characteristic of the lower mantle. The lower mantle is more or less solid, but the region is also very hot, and thus the rocks can flow very slowly (a process known as creep).

During the late 20th and early 21st centuries, scientific understanding of the deep mantle was greatly enhanced by high-resolution seismological studies combined with numerical modeling and laboratory experiments that mimicked conditions near the core-mantle boundary. Collectively, these studies revealed that the deep mantle is highly heterogeneous and that the layer may play a fundamental role in driving Earth's plates.

At a depth of about 2,900 km (1,800 miles), the lower mantle gives way to Earth's outer core, which is made up of a liquid rich in iron and nickel. At a depth of about 5,100 km (3,200 miles), the outer core transitions to the inner core. Although it has a higher temperature than the outer core, the inner core is solid because of the tremendous pressures that exist near Earth's centre. Earth's inner core is divided into the outer-inner core (OIC) and the inner-inner core (IIC), which differs from one another with respect to the polarity of their iron crystals. The polarity of the iron crystals of the OIC is oriented in a north-south direction, whereas that of the IIC is oriented east-west.

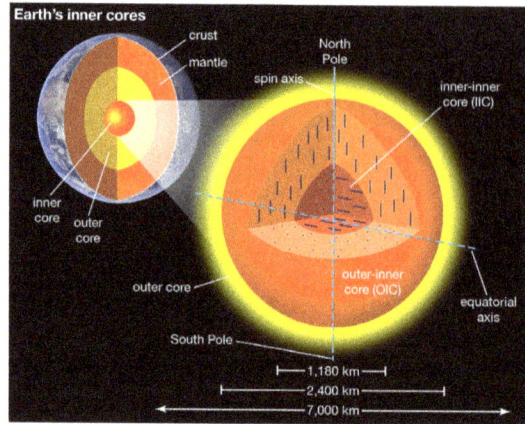

Earth's core: The internal layers of Earth's core, including its two inner cores.

Plate Boundaries

Lithospheric plates are much thicker than oceanic or continental crust. Their boundaries do not usually coincide with those between oceans and continents, and their behaviour is only partly influenced by whether they carry oceans, continents, or both. The Pacific Plate, for example, is entirely oceanic, whereas the North American Plate is capped by continental crust in the west (the North American continent) and by oceanic crust in the east and extends under the Atlantic Ocean as far as the Mid-Atlantic Ridge.

In a simplified example of plate motion shown in the figure, movement of plate A to the left relative to plates B and C results in several types of simultaneous interactions along the plate boundaries. At the rear, plates A and B move apart, or diverge, resulting in extension and the formation of a divergent margin. At the front, plates A and B overlap, or converge, resulting in compression and the formation of a convergent margin. Along the sides, the plates slide past one another, a process called shear. As these zones of shear link other plate boundaries to one another, they are called transform faults.

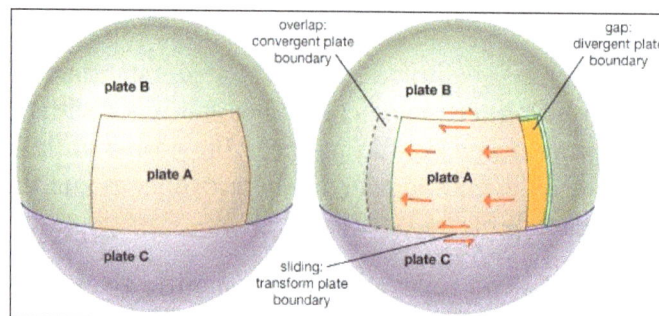

Theoretical diagram showing the effects of an advancing tectonic plate on other adjacent, but stationary, tectonic plates. At the advancing edge of plate A, the overlap with plate B creates a convergent boundary. In contrast, the gap left behind the trailing edge of plate A forms a divergent boundary with plate B. As plate A slides past portions of both plate B and plate C, transform boundaries develop.

Divergent Margins

As plates move apart at a divergent plate boundary, the release of pressure produces partial melting of the underlying mantle. This molten material, known as magma, is basaltic in composition and is buoyant. As a result, it wells up from below and cools close to the surface to generate new crust. Because new crust is formed, divergent margins are also called constructive margins.

Continental Rifting

Upwelling of magma causes the overlying lithosphere to uplift and stretch. (Whether magmatism [the formation of igneous rock from magma] initiates the rifting or whether rifting decompresses the mantle and initiates magmatism is a matter of significant debate.) If the diverging plates are capped by continental crust, fractures develop that are invaded by the ascending magma, prying the continents farther apart. Settling of the continental blocks creates a rift valley, such as the present-day East African Rift Valley. As the rift continues to widen, the continental crust becomes progressively thinner until separation of the plates is achieved and a new ocean is created. The ascending partial melt cools and crystallizes to form new crust. Because the partial melt is basaltic in composition, the new crust is oceanic, and an ocean ridge develops along the site of the former continental rift. Consequently, diverging plate boundaries, even if they originate within continents, eventually come to lie in ocean basins of their own making.

Rift Valley; Thingvellir National Park: The Thingvellir fracture zone at Thingvellir National Park in southwestern Iceland is an example of a rift valley. The Thingvellir fracture lies in the Mid-Atlantic Ridge, which extends through the centre of Iceland.

Seafloor Spreading

As upwelling of magma continues, the plates continue to diverge, a process known as seafloor spreading. Samples collected from the ocean floor show that the age of oceanic crust increases with distance from the spreading centre—important evidence in favour of this process. These age data also allow the rate of seafloor spreading to be determined, and they show that rates vary from about 0.1 cm (0.04 inch) per year to 17 cm (6.7 inches) per year. Seafloor-spreading rates are much more rapid in the Pacific Oceanthan in the Atlantic and Indian oceans. At spreading rates of about

15 cm (6 inches) per year, the entire crust beneath the Pacific Ocean (about 15,000 km [9,300 miles] wide) could be produced in 100 million years.

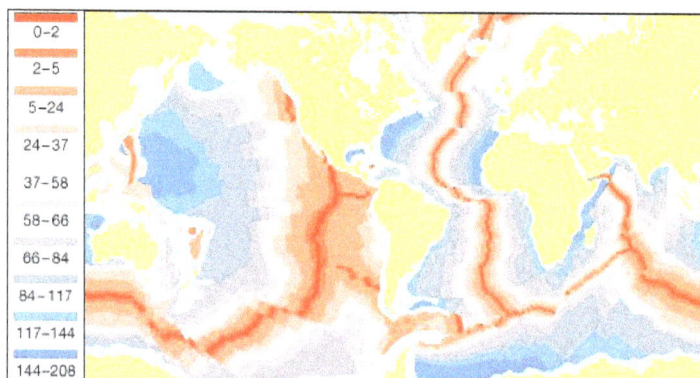

Age of Earth's oceanic crust: Map showing the age of Earth's oceanic crust and the pattern of seafloor spreading at the global scale.

Divergence and creation of oceanic crust are accompanied by much volcanic activity and by many shallow earthquakes as the crust repeatedly rifts, heals, and rifts again. Brittle earthquake-prone rocks occur only in the shallow crust. Deep earthquakes, in contrast, occur less frequently, due to the high heat flow in the mantle rock. These regions of oceanic crust are swollen with heat and so are elevated by 2 to 3 km (1.2 to 1.9 miles) above the surrounding seafloor. The elevated topography results in a feedback scenario in which the resulting gravitational force pushes the crust apart, allowing new magma to well up from below, which in turn sustains the elevated topography. Its summits are typically 1 to 5 km (0.6 to 3.1 miles) below the ocean surface. On a global scale, these ridges form an interconnected system of undersea "mountains" that are about 65,000 km (40,000 miles) in length and are called oceanic ridges.

Convergent Margins

Given that Earth is constant in volume, the continuous formation of Earth's new crust produces an excess that must be balanced by destruction of crust elsewhere. This is accomplished at convergent plate boundaries, also known as destructive plate boundaries, where one plate descends at an angle—that is, is subducted—beneath the other.

Because oceanic crust cools as it ages, it eventually becomes denser than the underlying asthenosphere, and so it has a tendency to subduct, or dive under, adjacent continental plates or younger sections of oceanic crust. The life span of the oceanic crust is prolonged by its rigidity, but eventually this resistance is overcome. Experiments show that the subducted oceanic lithosphere is denser than the surrounding mantle to a depth of at least 600 km (about 400 miles).

The mechanisms responsible for initiating subduction zones are controversial. During the late 20th and early 21st centuries, evidence emerged supporting the notion that subduction zones preferentially initiate along pre-existing fractures (such as transform faults) in the oceanic crust. Irrespective of the exact mechanism, the geologic record indicates that the resistance to subduction is overcome eventually.

Where two oceanic plates meet, the older, denser plate is preferentially subducted beneath the younger, warmer one. Where one of the plate margins is oceanic and the other is continental, the greater buoyancy of continental crust prevents it from sinking, and the oceanic plate is preferentially subducted. Continents are preferentially preserved in this manner relative to oceanic crust, which is continuously recycled into the mantle. This explains why ocean floor rocks are generally less than 200 million years old whereas the oldest continental rocks are more than 4 billion years old. Before the middle of the 20th century, most geoscientists maintained that continental crust was too buoyant to be subducted. However, it later became clear that slivers of continental crust adjacent to the deep-sea trench, as well as sediments deposited in the trench, may be dragged down the subduction zone. The recycling of this material is detected in the chemistry of volcanoes that erupt above the subduction zone.

Two plates carrying continental crust collide when the oceanic lithosphere between them has been eliminated. Eventually, subduction ceases and towering mountain ranges, such as the Himalayas, are created.

Because the plates form an integrated system, it is not necessary that new crust formed at any given divergent boundary be completely compensated at the nearest subduction zone, as long as the total amount of crust generated equals that destroyed.

Subduction Zones

The subduction process involves the descent into the mantle of a slab of cold hydrated oceanic lithosphere about 100 km (60 miles) thick that carries a relatively thin cap of oceanic sediments. The path of descent is defined by numerous earthquakes along a plane that is typically inclined between 30° and 60° into the mantle and is called the Wadati-Benioff zone, for Japanese seismologist Kiyoo Wadati and American seismologist Hugo Benioff, who pioneered its study. Between 10 and 20 percent of the subduction zones that dominate the circum-Pacific ocean basin are subhorizontal (that is, they subduct at angles between 0° and 20°). The factors that govern the dip of the subduction zone are not fully understood, but they probably include the age and thickness of the subducting oceanic lithosphere and the rate of plate convergence.

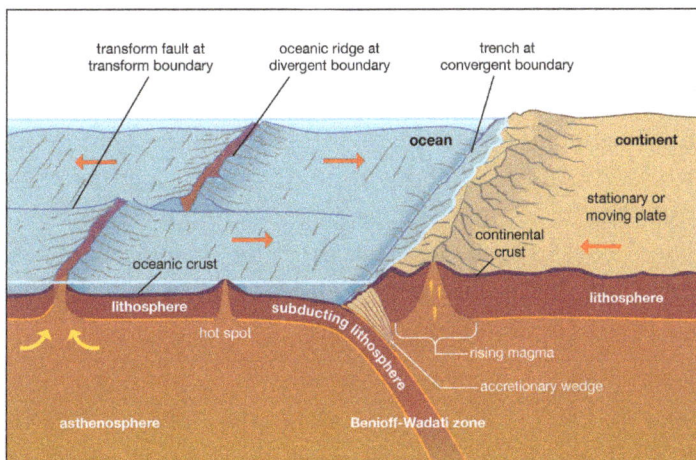

Subducting tectonic plate: A subducting plate's path (called the Benioff-Wadati [or Wadati-Benioff] zone) is defined by numerous earthquakes along a plane that is typically inclined between 30° and 60° into the mantle.

Most, but not all, earthquakes in this planar dipping zone result from compression, and the seismic activity extends 300 to 700 km (200 to 400 miles) below the surface, implying that the subducted crust retains some rigidity to this depth. At greater depths the subducted plate is partially recycled into the mantle.

The site of subduction is marked by a deep trench, between 5 and 11 km (3 and 7 miles) deep, that is produced by frictional drag between the plates as the descending plate bends before it subducts. The overriding plate scrapes sediments and elevated portions of ocean floor off the upper crust of the lower plate, creating a zone of highly deformed rocks within the trench that becomes attached, or accreted, to the overriding plate. This chaotic mixture is known as an accretionary wedge.

The rocks in the subduction zone experience high pressures but relatively low temperatures, an effect of the descent of the cold oceanic slab. Under these conditions the rocks recrystallize, or metamorphose, to form a suite of rocks known as blueschists, named for the diagnostic blue mineral called glaucophane, which is stable only at the high pressures and low temperatures found in subduction zones. At deeper levels in the subduction zone (that is, greater than 30–35 km [about 19–22 miles]), eclogites, which consist of high-pressure minerals such as red garnet (pyrope) and omphacite (pyroxene), form. The formation of eclogite from blueschist is accompanied by a significant increase in density and has been recognized as an important additional factor that facilitates the subduction process.

Island Arcs

When the downward-moving slab reaches a depth of about 100 km (60 miles), it gets sufficiently warm to drive off its most volatile components, thereby stimulating partial melting of mantle in the plate above the subduction zone (known as the mantle wedge). Melting in the mantle wedge produces magma, which is predominantly basaltic in composition. This magma rises to the surface and gives birth to a line of volcanoes in the overriding plate, known as a volcanic arc, typically a few hundred kilometres behind the oceanic trench. The distance between the trench and the arc, known as the arc-trench gap, depends on the angle of subduction. Steeper subduction zones have relatively narrow arc-trench gaps. A basin may form within this region, known as a fore-arc basin, and may be filled with sediments derived from the volcanic arc or with remains of oceanic crust.

If both plates are oceanic, as in the western Pacific Ocean, the volcanoes form a curved line of islands, known as an island arc, that is parallel to the trench, as in the case of the Mariana Islands and the adjacent Mariana Trench. If one plate is continental, the volcanoes form inland, as they do in the Andes of western South America. Though the process of magma generation is similar, the ascending magma may change its composition as it rises through the thick lid of continental crust, or it may provide sufficient heat to melt the crust. In either case, the composition of the volcanic mountains formed tends to be more silicon-rich and iron- and magnesium-poor relative to the volcanic rocks produced by ocean-ocean convergence.

Back-arc Basins

Where both converging plates are oceanic, the margin of the older oceanic crust will be subducted because older oceanic crust is colder and therefore denser. As the dense slab collapses into the asthenosphere, however, it also may "roll back" oceanward and cause extension in the overlying plate. This results in a process known as back-arc spreading, in which a basin opens up behind the island arc. The crust behind the arc becomes progressively thinner, and the decompression of the underlying mantle causes the crust to melt, initiating seafloor-spreading processes, such as melting and the production of basalt; these processes are similar to those that occur at ocean ridges. The geochemistry of the basalts produced at back-arc basins superficially resembles that of basalts produced at ocean ridges, but subtle trace element analyses can detect the influence of a nearby subducted slab.

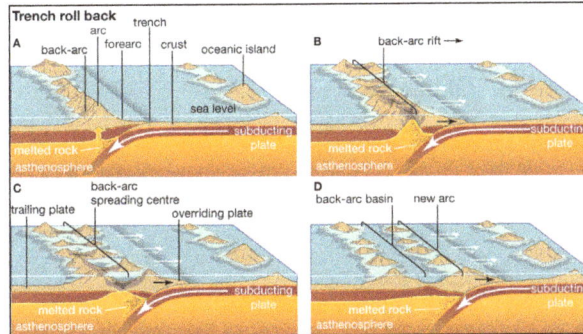

The trench "rolls back" process of back-arc basin formation.

This style of subduction predominates in the western Pacific Ocean, in which a number of back-arc basins separate several island arcs from Asia. Examples include the Mariana Islands, the Kuril Islands, and the main islands of Japan. However, if the rate of convergence increases or if anomalously thick oceanic crust (possibly caused by rising mantle plume activity) is conveyed into the subduction zone, the slab may flatten. Such flattening causes the back-arc basin to close, resulting in deformation, metamorphism, and even melting of the strata deposited in the basin.

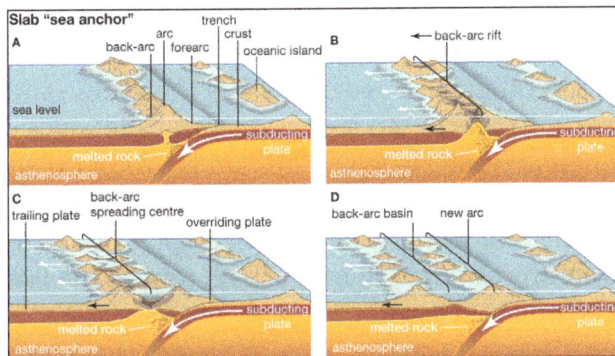

The slab "sea anchor" process of back-arc basin formation.

Mountain Building

If the rate of subduction in an ocean basin exceeds the rate at which the crust is formed at oceanic ridges, a convergent margin forms as the ocean initially contracts. This process can lead to collision between the approaching continents, which eventually terminates subduction. Mountain building can occur in a number of ways at a convergent margin: mountains may rise as a consequence of the subduction process itself, by the accretion of small crustal fragments (which, along with linear island chains and oceanic ridges, are known as terranes), or by the collision of two large continents.

Many mountain belts were developed by a combination of these processes. For example, the Cordilleran mountain belt of North America—which includes the Rocky Mountains as well as the Cascades, the Sierra Nevada, and other mountain ranges near the Pacific coast—developed by a

combination of subduction and terrane accretion. As continental collisions are usually preceded by a long history of subduction and terrane accretion, many mountain belts record all three
processes. Over the past 70 million years the subduction of the Neo-Tethys Sea, a wedge-shaped
body of water that was located between Gondwana and Laurasia, led to the accretion of terranes
along the margins of Laurasia, followed by continental collisions beginning about 30 million years
ago between Africa and Europe and between India and Asia. These collisions culminated in the
formation of the Alps and the Himalayas.

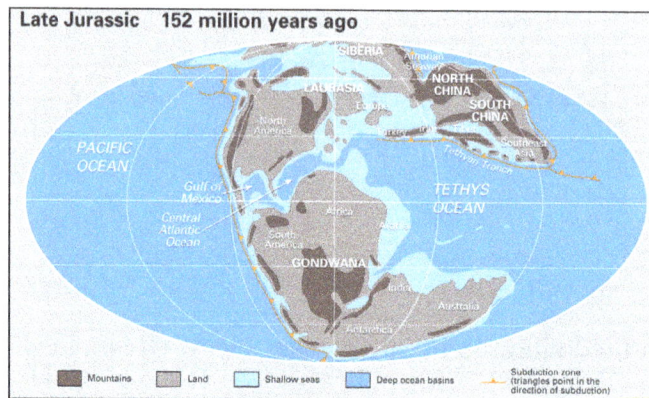

Jurassic paleogeography: Distribution of landmasses, mountainous regions, shallow seas, and deep ocean
basins during the late Jurassic Period. Included in the paleogeographic reconstruction
are the locations of the interval's subduction zones.

Mountains by Subduction

Mountain building by subduction is classically demonstrated in the Andes Mountains of South
America. Subduction results in voluminous magmatism in the mantle and crust overlying the subduction zone, and, therefore, the rocks in this region are warm and weak. Although subduction is a
long-term process, the uplift that results in mountains tends to occur in discrete episodes and may
reflect intervals of stronger plate convergence that squeezes the thermally weakened crust upward.
For example, rapid uplift of the Andes approximately 25 million years ago is evidenced by a rever-
sal in the flow of the Amazon River from its ancestral path toward the Pacific Ocean to its modern
path, which empties into the Atlantic Ocean.

In addition, models have indicated that the episodic opening and closing of back-arc basins have
been the major factors in mountain-building processes, which have influenced the plate-tectonic
evolution of the western Pacific for at least the past 500 million years.

Mountains by Terrane Accretion

As the ocean contracts by subduction, elevated regions within the ocean basin—terranes—
are transported toward the subduction zone, where they are scraped off the descending
plate and added—accreted—to the continental margin. Since the late Devonian and early
Carboniferousperiods, some 360 million years ago, subduction beneath the western margin of
North America has resulted in several collisions with terranes. The piecemeal addition of these
accreted terranes has added an average of 600 km (400 miles) in width along the western
margin of the North American continent, and the collisions have resulted in important pulses
of mountain building.

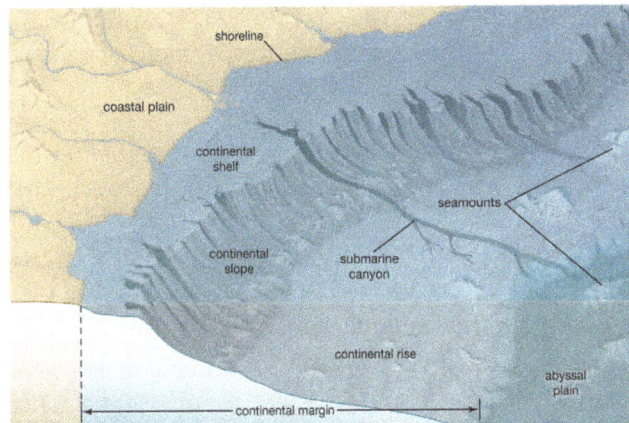

Continental margin: The broad, gentle pitch of the continental shelf gives way to the relatively steep continental slope. The more gradual transition to the abyssal plain is a sediment-filled region called the continental rise. The continental shelf, slope, and rise are collectively called the continental margin.

During these accretionary events, small sections of the oceanic crust may break away from the subducting slab as it descends. Instead of being subducted, these slices are thrust over the overriding plate and are said to be obducted. Where this occurs, rare slices of ocean crust, known as ophiolites, are preserved on land. They provide a valuable natural laboratory for studying the composition and character of the oceanic crust and the mechanisms of their emplacement and preservation on land. A classic example is the Coast Range ophiolite of California, which is one of the most extensive ophiolite terranes in North America. These ophiolite deposits run from the Klamath Mountains in northern California southward to the Diablo Range in central California. This oceanic crust likely formed during the middle of the Jurassic Period, roughly 170 million years ago, in an extensional regime within either a back-arc or a forearc basin. In the late Mesozoic, it was accreted to the western North American continental margin.

Because preservation of oceanic crust is rare, the recognition of ophiolite complexes is very important in tectonic analyses. Until the mid-1980s, ophiolites were thought to represent vestiges of the main oceanic tract, but geochemical analyses have clearly indicated that most ophiolites form near volcanic arcs, such as in back-arc basins characterized by subduction roll-back (the collapse of the subducting plate that causes the extension of the overlying plate). The recognition of ophiolite complexes is very important in tectonic analysis, because they provide insights into the generation of magmatism in oceanic domains, as well as their complex relationships with subduction processes.

Mountains by Continental Collision

Continental collision involves the forced convergence of two buoyant plate margins that results in neither continent being subducted to any appreciable extent. A complex sequence of events ensues that compels one continent to override the other. These processes result in crustal thickening and intense deformation that forces the crust skyward to form huge mountains with crustal roots that extend as deep as 80 km (about 50 miles) relative to Earth's surface, in accordance with the principles of isostasy.

The subducted slab still has a tendency to sink and may become detached and founder (submerge) into the mantle. The crustal root undergoes metamorphic reactions that result in a significant increase in density and may cause the root to also founder into the mantle. Both processes result in a significant injection of heat from the compensatory upwelling of asthenosphere, which is an important contribution to the rise of the mountains.

Continental collisions produce lofty landlocked mountain ranges such as the Himalayas. Much later, after these ranges have been largely leveled by erosion, it is possible that the original contact, or suture, may be exposed.

The balance between creation and destruction on a global scale is demonstrated by the expansion of the Atlantic Ocean by seafloor spreadingover the past 200 million years, compensated by the contraction of the Pacific Ocean, and the consumption of an entire ocean between India and Asia (the Tethys Sea). The northward migration of India led to collision with Asia some 40 million years ago. Since that time India has advanced a further 2,000 km (1,250 miles) beneath Asia, pushing up the Himalayas and forming the Plateau of Tibet. Pinned against stable Siberia, China and Indo-china were pushed sideways, resulting in strong seismic activity thousands of kilometres from the site of the continental collision.

Transform Faults

Along the third type of plate boundary, two plates move laterally and pass each other along giant fractures in Earth's crust. Transform faults are so named because they are linked to other types of plate boundaries. The majority of transform faults link the offset segments of oceanic ridges. However, transform faults also occur between plate margins with continental crust—for example, the San Andreas Fault in California and the North Anatolian fault system in Turkey. These boundaries are conservativebecause plate interaction occurs without creating or destroying crust. Because the only motion along these faults is the sliding of plates past each other, the horizontal direction along the fault surface must parallel the direction of plate motion. The fault surfaces are rarely smooth, and pressure may build up when the plates on either side temporarily lock. This buildup of stress may be suddenly released in the form of an earthquake.

Section of the San Andreas Fault.

Many transform faults in the Atlantic Ocean are the continuation of major faults in adjacent continents, which suggests that the orientation of these faults might be inherited from pre-existing weaknesses in continental crust during the earliest stages of the development of oceanic crust. On the other hand, transform faults may themselves be reactivated, and recent geodynamic models suggest that they are favourable environments for the initiation of subduction zones.

Hotspots

Although most of Earth's volcanic activity is concentrated along or adjacent to plate boundaries, there are some important exceptions in which this activity occurs within plates. Linear chains of islands, thousands of kilometres in length, that occur far from plate boundaries are the most notable examples. These island chains record a typical sequence of decreasing elevation along the chain, from volcanic island to fringing reefto atoll and finally to submerged seamount. An active volcano usually exists at one end of an island chain, with progressively older extinct volcanoes occurring along the rest of the chain. Canadian geophysicist J. Tuzo Wilsonand American geophysicist W. Jason Morgan explained such topographic features as the result of hotspots.

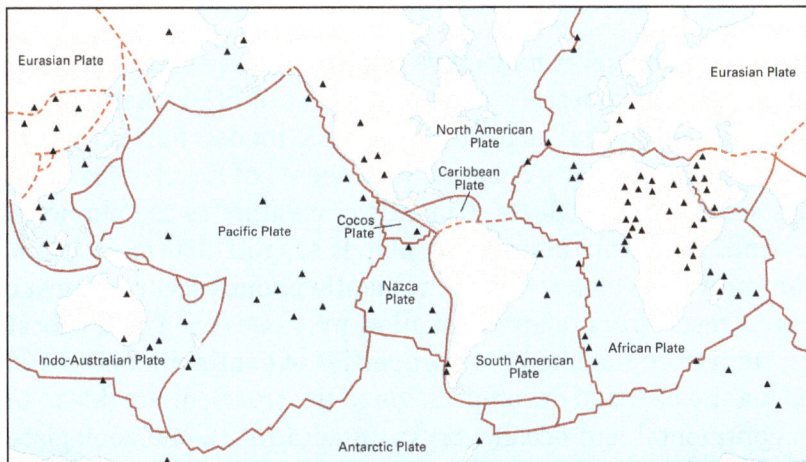

The principal tectonic plates that make up Earth's lithosphere. Also located are several dozen hot spots where plumes of hot mantle material are upwelling beneath the plates.

The number of these hotspots is uncertain (estimates range from 20 to 120), but most occur within a plate rather than at a plate boundary. Hotspots are thought to be the surface expression of giant plumes of heat, termed mantle plumes, that ascend from deep within the mantle, possibly from the core-mantle boundary, some 2,900 km (1,800 miles) below the surface. These plumes are thought to be stationary relative to the lithospheric plates that move over them. A volcano builds upon the surface of a plate directly above the plume. As the plate moves on, however, the volcano is separated from its underlying magma source and becomes extinct. Extinct volcanoes are eroded as they cool and subside to form fringing reefs and atolls, and eventually they sink below the surface of the sea to form a seamount. At the same time, a new active volcano forms directly above the mantle plume.

The best example of this process is preserved in the Hawaiian-Emperor seamount chain. The plume is presently situated beneath Hawaii, and a linear chain of islands, atolls, and seamounts extends 3,500 km (2,200 miles) northwest to Midway and a further 2,500 km (1,500 miles) north-north-west to the Aleutian Trench. The age at which volcanism became extinct along this chain gets progressively older with increasing distance from Hawaii—critical evidence that supports this theory. Hotspot volcanism is not restricted to the ocean basins; it also occurs within continents, as in the case of Yellowstone National Park in western North America.

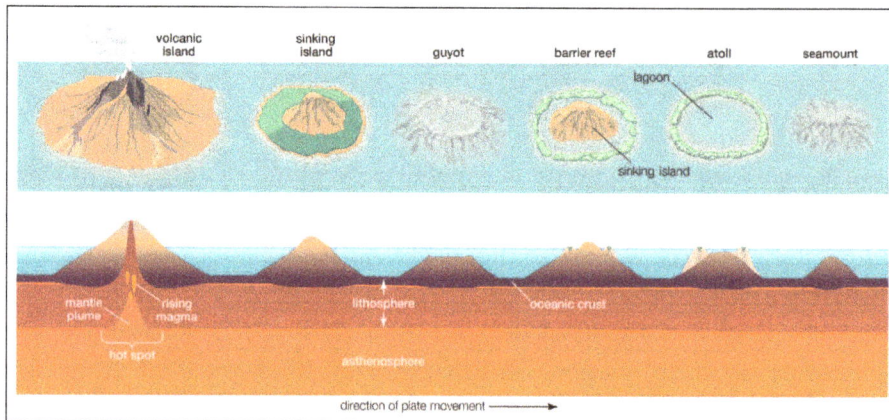

Diagram depicting the process of atoll formation. Atolls are formed from the remnant parts of sinking volcanic islands.

Measurements suggest that hotspots may move relative to one another, a situation not predicted by the classical model, which describes the movement of lithospheric plates over stationary mantle plumes. This has led to challenges to this classic model. Furthermore, the relationship between hotspots and plumes is hotly debated. Proponents of the classical model maintain that these discrepancies are due to the effects of mantle circulation as the plumes ascend, a process called the mantle wind. Data from alternative models suggest that many plumes are not deep-rooted. Instead, they provide evidence that many mantle plumes occur as linear chains that inject magma into fractures, result from relatively shallow processes such as the localized presence of water-rich mantle, stem from the insulating properties of continental crust (which leads to the buildup of trapped mantle heat and decompression of the crust), or are due to instabilities in the interface between continental and oceanic crust. In addition, some geologists note that many geologic processes that others attribute to the behaviour of mantle plumes may be explained by other forces.

Plate Motion

Euler's Contributions

Swiss mathematician Leonhard Euler showed that the movement of a rigid body across the surface of a sphere can be described as a rotation (or turning) around an axis that goes through the centre of the sphere, known as the axis of rotation. The location of this axis bears no relationship to Earth's spin axis. The point of emergence of the axis through the surface of the sphere is known as the pole of rotation. This theorem of spherical geometry provides an elegant way to define the motion of the lithospheric plates across Earth's surface. Therefore, the relative motion of two rigid plates may be described as rotations around a common axis, known as the axis of spreading. Application of the theorem requires that the plates not be internally deformed—a requirement not absolutely adhered to but one that appears to be a reasonable approximation of what actually happens. Application of this theorem permits the mathematical reconstruction of past plate configurations.

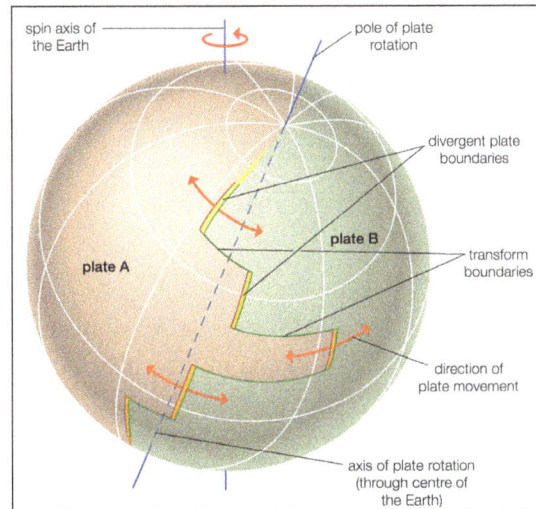

Theoretical depiction of the movement of tectonic plates across Earth's surface. Movement on a sphere of two plates, A and B, can be described as a rotation around a common pole. Circles around that pole correspond to the orientation of transform faults (that is, single lines in the horizontal that connect to divergent plate boundaries, marked by double lines, in the vertical).

Because all plates form a closed system, all movements can be defined by dealing with them two at a time. The joint pole of rotation of two plates can be determined from their transform boundaries, which are by definition parallel to the direction of motion. Thus, the plates move along transform faults, whose trace defines circles of latitude perpendicular to the axis of spreading, and so form small circles around the pole of rotation. A geometric necessity of this theorem—that lines perpendicular to the transform faults converge on the pole of rotation—is confirmed by measurements. According to this theorem, the rate of plate motion should be slowest near the pole of rotation and increase progressively to a maximum rate along fractures with a 90° angle to it. This relationship is also confirmed by accurate measurements of sea-floor-spreading rates.

Past Plate Movements

Plate tectonics involves the movements of Earth's lithospheric plates relative to one another over the planet's weak asthenosphere. This activity changes the positions of all plates with respect to Earth's spin axis and the Equator. To determine the true geographic positions of the plates in the past, investigators have to define their motions, not only relative to each other but also relative to this independent frame of reference. Hotspots, as classically interpreted, provide an example of such a reference frame, assuming they are the sources of plumes that originate within the deep mantle and have relatively fixed positions over time. If this assumption is valid, the motion of the lithosphere above these plumes can be deduced. The hotspot island chains serve this purpose, their trends providing the direction of motion of a plate. The speed of the plate can be inferred from the increase in age of the volcanoes along the chain relative to the distance between the islands.

Earth scientists are able to accurately reconstruct the positions and movements of plates for the past 150 million to 200 million years because they have the oceanic crust record to provide them with plate speeds and direction of movement. However, since older oceanic crust is continuously consumed to make room for new crust, this kind of evidence is not available for earlier intervals of geologic time, making it necessary for investigators to turn to other, less-precise techniques.

Interactions of Tectonics with other Systems

Oceans-sea Level

As plate tectonics changes the shape of ocean basins, it fundamentally affects long-term variations in global sea level. For example, the geologic record in which thick sequences of continental shelf sediments were deposited demonstrates that the breakup of Pangea resulted in the flooding of continental margins, indicating a rise in sea level. There are several contributing factors. First, the presence of new ocean ridges displaces seawater upward and outward across the continental margins. Second, the dispersing continental fragments subside as they cool. Third, the volcanism associated with breakup introduces greenhouse gases in the atmosphere, which results in global warming, causing continental glaciers to melt.

As an ocean widens, its crust becomes older and denser. It therefore subsides, eventually forming ocean trenches. As a result, ocean basins can hold more water, and sea level drops. This changes once again when subduction commences. Subduction preferentially consumes the oldest oceanic crust, so that the average age of oceanic crust becomes younger. Younger oceanic crust is therefore more buoyant and has a higher elevation, a circumstance that causes sea level to rise once more.

Composition of Ocean Water

Water's strong properties as a solvent mean that it is rarely pure. Ocean water contains about 96.5 percent by weight pure water, with the remaining 3.5 percent predominantly consisting of ions such as chloride (1.9 percent), sodium, (1.1 percent), sulfate (0.3 percent), and magnesium (0.1 percent). The drainage of water from continents via the hydrologic cycle-plays an important role in transporting chemicals from the land to the sea. The effect of this drainage is profoundly influenced by the presence of mountain belts. For example, the erosional power of the Ganges River, which drains from the Himalayas, carries 1.45 billion metric tons of sediment to the sea annually. This load is nine times that of the Mississippi River. The processes of weathering and erosion strip soluble elements such as sodium from their host minerals, and the relatively high concentration of sodium in ocean water is attributed to the weathering and erosion that accompanies continental drainage.

Until the advent of plate tectonics, discovering the source of chlorine was problematic because chlorine is present in only very minor amounts in the continental crust. Scientists hypothesized that the source of this element may lie in underwater volcanic activity. In the late 1970s three scientists investigating the oceanic ridge off the coast of Peru from a submersible craft documented the occurrence of superheated jets of water, up to 350 °C (660 °F), continuously erupting from chimneys that stood 13 metres (43 feet) above the ocean floor. These hot springs were found to be rich in chlorine and metals, confirming that the source of chlorine in the oceans lay in the tectonic processes occurring at oceanic ridges.

Life

Plate tectonics has influenced the evolution and propagation of life in a variety of ways. The study of oceanic ridges revealed the presence of bizarre life adjacent to the chimneys of superheated water that together makes up about 1 percent of the world's ecosystems. The existence of these life-forms in the deep ocean cannot be based on photosynthesis. Instead, they are nourished by minerals and heat. The energy released when hydrogen sulfide in the vent reacts with seawater is used by bacteria to convert inorganic carbon dioxide dissolved in seawater into organic compounds, a process known as chemosynthesis. Some scientists speculate that the cumulative influence of this process over time has had a significant effect on evolution. Others suggest that similar processes may ultimately be responsible for the origin of life on Earth.

Vent tube worms (Riftia pachyptila).

The continuous rearrangement of the size and shape of ocean basins and continents over geologic time, accompanied by changes in ocean circulation and climate, had a major impact on the development of life on Earth. One of the first studies of the potential effects of plate tectonics on life was published in 1970 by American geologists James W. Valentine and Eldridge M. Moores, who proposed that the diversity of life increased as continents fragmented and dispersed and diversity diminished when the continents were joined together.

Evolution

Opossum: Nocturnal animals, such as opossums, have eyes with large, nearly spherical lenses.

When Pangea began to rift apart and the Atlantic Ocean started to open during the middle Mesozoic, the differences between the faunas of opposite shores gradually increased in an almost linear fashion—the greater the distance, the smaller the number of families in common. The difference increased more rapidly in the South Atlantic than in the North Atlantic, where a land connection between Europe and North Americapersisted until about 60 million years ago.

After the breakup of Pangea, no land animal or group of animals could become dominant, because the continents were disconnected. As a result, the separated landmasses evolved highly specialized fauna. South America, for example, was rich in marsupial mammals, which had few predators. North America, on the other hand, was rich in placental mammals.

However, about 3 million years ago, volcanic activity associated with subduction of the eastern Pacific Ocean formed a land bridge across the isthmus of Panama, reconnecting the separate landmasses. The emergence of the isthmus made it possible for land animals to cross, forcing previously separated fauna to compete. Numerous placental mammals and herbivores migrated from north to south. They adapted well to the new environment and were more successful than the local fauna in competing for food. The invasion of highly adaptable carnivores from the north contributed to the extinction of at least four orders of South American land mammals. A few species, notably the armadillo and the opossum, managed to migrate in the opposite direction. Ironically, many of the invading northerners, such as the llama and tapir, subsequently became extinct in their region of origin and found their last refuge to the south.

Extinction

Perhaps the most dramatic example of the potential impact of plate tectonics on life occurred near the end of the Permian Period (roughly 299 million to 252 million years ago). Several events contributed to the Permian extinction that caused the permanent disappearance of half of Earth's known biological families. The marine realm was most affected, losing more than 90 percent of its species. About 70 percent of terrestrial species became extinct. This extinction appears to have occurred in several pulses, and there may have been numerous contributing factors—including biogeographic changes associated with the formation of Pangea (which would have been accompanied by a sharp decrease in area of shallow-water habitats), changes in the patterns of nutrient-rich deep ocean currents, changes in the amount of dissolved oxygen in ocean waters, and temperature increases and changes to the carbon cycle caused partly by the population explosion of the methane-producing microbe *Methanosarcina*. Another contributing factor could have been the environmental consequences of the vast volcanic outpourings of the Siberian Traps, one of the largest volcanic events documented. It produced a region of flood basalt that had an estimated volume of 2,000,000–3,000,000 million cubic km [about 480,000–720,000 cubic miles]). The Siberian Traps eruption occurred about the same time as the extinction, and the greenhouse gases emitted from these volcanoes may have affected the amount of acidity of the oceans.

The extinction had a complex history. High latitudes were affected first as a result of the waning of the Permian ice age when the southern edge of Pangea moved off the South Pole. The equatorial and subtropical zones appear to have been affected somewhat later by a global cooling. On the other hand, the extinctions were not felt as strongly on the continent itself. Instead, the vast semi-arid and arid lands that emerged on so large a continent, the shortening of its moist coasts, and the many mountainranges formed from the collisions that led to the formation of the supercontinent provided strong incentives for evolutionary adaptation to dry or high-altitude environments.

Climate

Climate changes associated with the supercontinent of Pangea and with its eventual breakup and dispersal provide an example of the effect of plate tectonics on paleoclimate. Pangea was completely surrounded by a world ocean (Panthalassa) extending from pole to pole and spanning 80 percent of the circumference of Earth at the paleoequator. The equatorial current system, driven by the trade winds, resided in warm latitudes much longer than today, and its waters were therefore warmer. The gyres that occupy most of the Southern and Northern hemispheres were also warmer, and consequently the temperature gradient from the paleoequator to the poles was less pronounced than it is at present.

Early in the Mesozoic Era, Gondwana split from its northern counterpart, Laurasia, to form the Tethys seaway, and the equatorial current became circumglobal. Equatorial surface waters were then able to circumnavigate the world and became even warmer. How this flow influenced circulation at higher latitudes is unclear. From about 100 million to 70 million years ago, isotopic records show, Arctic and Antarctic surface water temperatures were at or above 10 °C (50 °F), and the polar regions were warm enough to support forests.

As the dispersal of continents following the breakup of Pangea continued, however, the surface circulation of the oceans began to approach the more complex circulation patterns of today. About 100 million years ago, the northward drift of Australia and South America created a new circumglobal seaway around Antarctica, which remained centred on the South Pole. A vigorous circum-Antarctic current developed, isolating the southern continent from the warmer waters to the north. At the same time, the equatorial current system became blocked, first in the Indo-Pacific region and next in the Middle East and eastern Mediterranean and, about 6 million years ago, by the emergence of the Isthmus of Panama. As a result, the equatorial waters were heated less, and the midlatitude ocean gyres were not as effective in keeping the high-latitude waters warm. Because of this, an ice cap began to form on Antarctica some 20 million years ago and grew to roughly its present size about 5 million years later. This ice cap cooled the waters of the adjacent ocean to such a low temperature that the waters sank and initiated the north-directed abyssal flow that marks the present deep circulation.

Also at about 6 million years ago, the collision between Africa and Europetemporarily closed the Strait of Gibraltar, isolating the Mediterranean Seaand restricting its circulation. Evaporation, which produced thick salt deposits, virtually dried up this sea and lowered the salt content of the world's oceans, allowing seawater to freeze at higher temperatures. As a result, polar ice sheets grew and sea level fell. About 500,000 years later, the barrier between the Mediterranean and the Atlantic Ocean was breached, and open circulation resumed.

The Quaternary Ice Age arrived in full when the first ice caps appeared in the Northern Hemisphere about 2 million years ago. It is highly unlikely that the changing configuration of continents and oceans can be held solely responsible for the onset of the Quaternary Ice Age, even if such factors as the drift of continents across the latitudes (with the associated changes in vegetation) and reflectivity (albedo) for solar heat are included. There can be little doubt, however, that it was a contributing factor and that recognition of its role has profoundly altered concepts of paleoclimatology.

Tectonic Landforms

There are two basic forms of tectonic activity: compression and extension.

Compression occurs when lithospheric plates are squeezed together along converging lithospheric plate boundaries, while extension happens along continental and oceanic rifting, where plates are being pulled apart.

Fold Belts

When two continental lithospheric plates collide, the plates are squeezed together at the boundary. The crust crumples, creating folds. The Jura Mountains of France and Switzerland, shown in figure, are an example of relatively young—geologically speaking—open folds. The folds create a set of alternating anticlines, or up-arching bends, and troughs, called synclines.

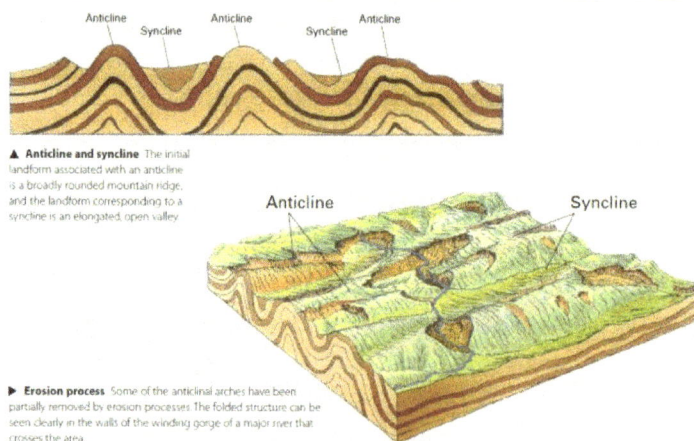

▲ **Anticline and syncline** The initial landform associated with an anticline is a broadly rounded mountain ridge, and the landform corresponding to a syncline is an elongated, open valley.

▶ **Erosion process** Some of the anticlinal arches have been partially removed by erosion processes. The folded structure can be seen clearly in the walls of the winding gorge of a major river that crosses the area.

Folds of The Jura Mountains.

When fold belts are eroded, they create a ridge-and-valley landscape, like the eastern side of the Appalachian Mountains from Pennsylvania to Alabama. In this landscape, weaker formations such as shale and limestone are eroded away, leaving hard strata, such as sandstone or quartzite, to stand in bold relief as long, narrow ridges. These folds are continuous and even-crested, producing almost parallel ridges. In some regions, however, the fold crests plunge, rising up in places and dipping down in others. This provides a pattern of curving mountain crests and valleys.

Note that anticlines are not always ridges. If a resistant rock type at the center of the anticline is eroded through to reveal softer rocks underneath, an anticlinal valley may form. A synclinal mountain is also possible. It occurs when a resistant rock type is exposed at the center of a syncline, and the rock forms a ridge.

Faults and Fault Landforms

A fault is a fracture created in the brittle rocks of the Earth's crust, as different parts of the crust move in different directions. Fault lines can sometimes be followed along the ground for many kilometers. Most major faults extend down into the crust for at least several kilometers.

Faults are evidence of relative movement between the rock on either side of the fault. Rock on either side suddenly slips along the fault plane, generating earthquakes. A single fault movement can cause slippage of as little as a centimeter or as much as 15 m (about 50 ft).

These movements typically happen many years or decades apart, or even several centuries apart. But when we add up all these small motions over long time spans, they can amount to tens or hundreds of kilometers. There are four main types of faults: normal, transcurrent, reverse, and overthrust faults.

Normal faults are a common type of fault produced by crustal rifting. They usually occur as a set of parallel faults creating fault scarps, grabens, and horsts. Where normal faulting occurs on a grand scale, it produces ranges of Block Mountains flanked by down dropped lowland basins.

When lithospheric plates slide past one another horizontally along major transform faults, we refer to these faults as transcurrent faults, or strike-slip faults. The San Andreas fault is a famous active transcurrent fault.

You can follow it for about 1000 km (about 600 mi) from the Gulf of California to Cape Mendocino.

Throughout many kilometers of its length, the San Andreas fault appears as a straight, narrow scar. In some places this scar is a trench-like feature, and elsewhere it is a low scarp.

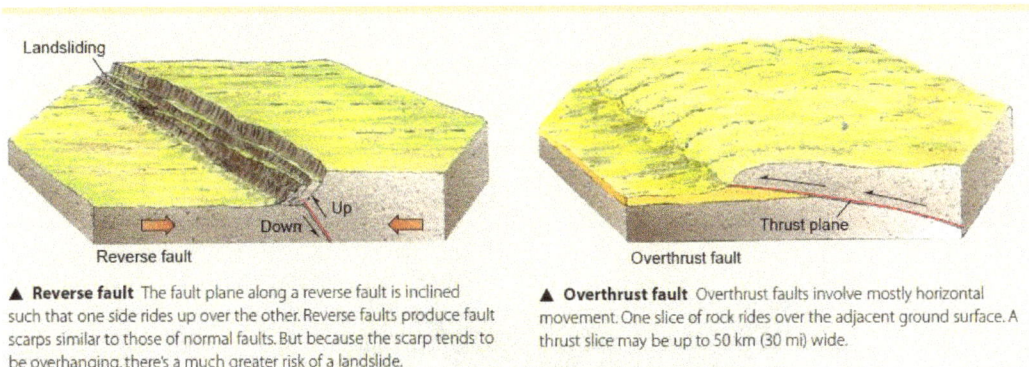

▲ **Reverse fault** The fault plane along a reverse fault is inclined such that one side rides up over the other. Reverse faults produce fault scarps similar to those of normal faults. But because the scarp tends to be overhanging, there's a much greater risk of a landslide.

▲ **Overthrust fault** Overthrust faults involve mostly horizontal movement. One slice of rock rides over the adjacent ground surface. A thrust slice may be up to 50 km (30 mi) wide.

Reverse and overthrust faults.

Figure illustrates the reverse fault and the overthrust fault. Both are caused by crustal compression. The San Fernando, California, earthquake of 1971 was generated by slippage on a reverse fault. When compression is severe, for example, in a continent-continent collision, rock layers can ride over each other on a low-angle overthrust fault. Repeated faulting can produce a great rock cliff hundreds of meters high. Because fault planes extend hundreds of meters down into the bedrock, their landforms can persist even after several million years of erosion. Figure diagrams the effect of erosion on a fault scarp.

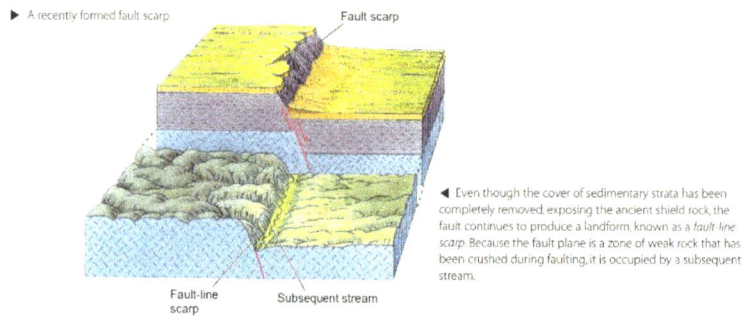

▶ A recently formed fault scarp

Fault scarp

◀ Even though the cover of sedimentary strata has been completely removed, exposing the ancient shield rock, the fault continues to produce a landform, known as a *fault-line scarp*. Because the fault plane is a zone of weak rock that has been crushed during faulting, it is occupied by a subsequent stream.

Fault-line scarp

Subsequent stream

Fault scarp evolution.

The East African Rift Valley System

The Rift Valley of East Africa provides an example of extension and normal faulting at a continental scale. Here, continental lithosphere is beginning to rupture and split apart in the first stage of forming a new ocean basin. The Rift Valley is basically a series of linked and branching grabens, but with a more complex history that includes building volcanoes on the graben floor.

Figure shows a map of the East African Rift Valley system. It is about 3000 km (1900 mi) long and extends from the Red Sea southward to Lake Nyassa. Along this axis, the Earth's crust is being lifted and spread apart. The rift valleys are like keystone blocks of a masonry arch that have slipped down between neighboring blocks because the arch has spread apart somewhat. Thus, the floors of the rift valleys are above the elevation of most of the African continental surface. Major rivers and several long, deep lakes—Lake Nyassa and Lake Turkana, for example—occupy some of the valley floors.

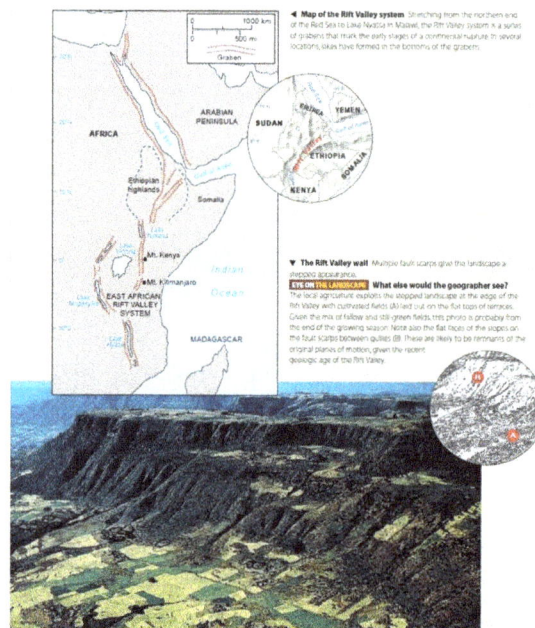

The sides of the rift valleys typically consist of fault scarps and multiple fault steps. Sediments from the adjacent plateaus make thick fills in the floors of the valleys. Two great strato-volcanoes have been built close to the Rift Valley east of Lake Victoria. One is Mount Kilimanjaro, whose summit rises to over 6000 m (about 19,000 ft). The other, Mount Kenya, is only a little lower and lies right on the Equator.

Supercontinents

The motion of the tectonic plates periodically causes most of the continental landmasses of the planet to collide with each other, forming giant continents known as supercontinents. For much of the past several billion years, these supercontinents have alternately formed and broken up in a process called the supercontinent cycle. The last supercontinent was known as Pangaea, which broke up about 160 million years ago to form the present-day plates on the planet. Before that the previous supercontinent was Gondwana, which formed about 600-500 million years ago, and the one before that was Rodinia, formed around a billion years ago. The distribution of landmasses and formation and breakup of supercontinents has dramatically influenced global and local climate on timescales of 100 million years, with cycles repeating for the past few billion years of Earth's history. The supercontinent cycle predicts that the planet should have periods of global warming associated with supercontinent breakup, and global cooling associated with supercontinent formation. The supercontinent cycle affects sea-level changes, initiates periods of global glaciation, and changes the global climate from hothouse to icehouse conditions, and influences seawater salinity and nutrient supply. All of these consequences of plate tectonics have profound influences on life on Earth.

Sea level has changed by thousands of feet (hundreds of meters) above and below current levels many times during Earth's history. In fact, sea level is constantly changing in response to a number of different variables, many of them related to plate tectonics, the supercontinent cycle, and climate. Sea level was 1,970 feet (600 m) higher than now during the ordovician and reached a low stand at the end of the Permian. Sea levels were high again in the Cretaceous during the breakup of the supercontinent of Pangaea.

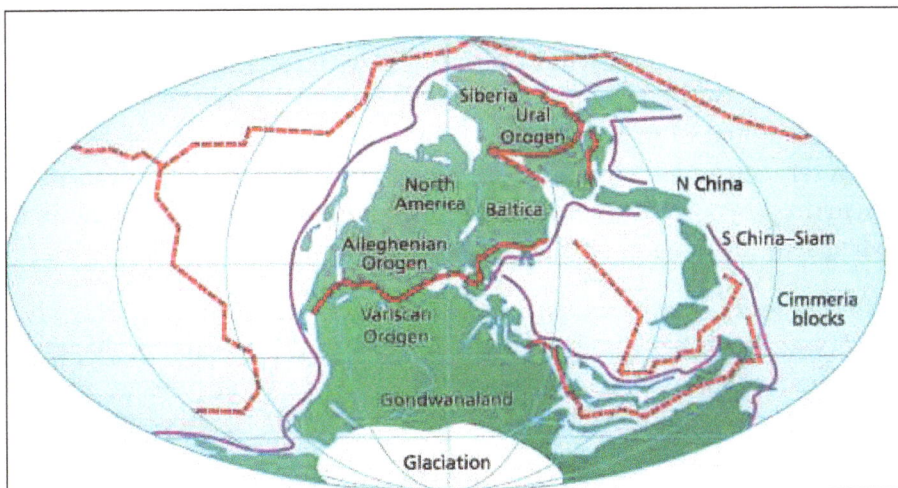

Pangaea Supercontinent Formation

Late Carboniferous 300 Ma Global icehouse; low sea level; continental collisions.

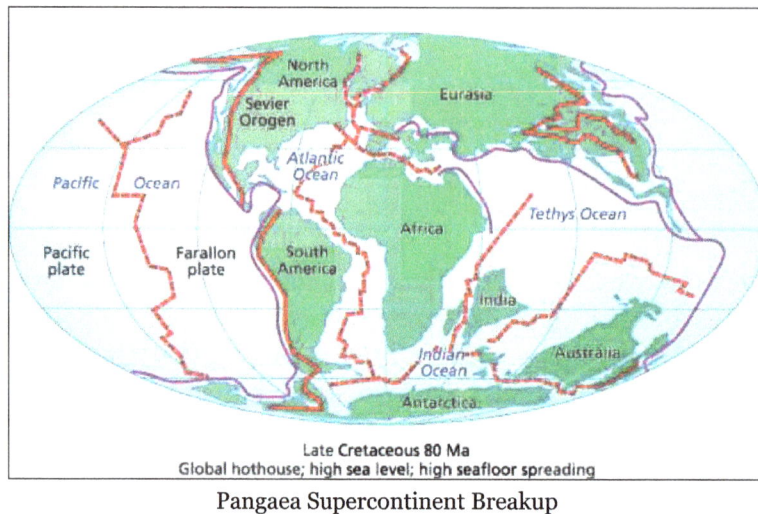

Late Cretaceous 80 Ma
Global hothouse; high sea level; high seafloor spreading

Pangaea Supercontinent Breakup

Maps of continental positions during cold and warm climates showing the relationship between climate and tectonics sea levels may change at different rates and amounts in response to different phases of the supercontinent cycle, and the sea level changes are closely related to climate. The global volume of the midocean ridges can change dramatically, either by increasing the total length of ridges or changing the rate of seafloor spreading. Either process produces more volcanism; increases the volume of volcanoes on the seafloor, raising sea levels; and puts a lot of extra CO_2 into the atmosphere, raising global temperatures. The total length of ridges typically increases during continental breakup, since continents are being rifted apart and some continental rifts can evolve into midocean ridges. Additionally, if seafloor spreading rates are increased, the amount of young, topographically elevated ridges is increased relative to the slower, older topographically lower ridges that occupy a smaller volume. If the volume of the ridges increases by either mechanism, then a volume of water equal to the increased ridge volume is displaced and sea level rises, inundating the continents. Changes in ridge volume are able to change sea levels positively or negatively by about 985 feet (300 m) from present values, at rates of about 0.4 inch (1 cm) every 1,000 years.

Continent-continent collisions, such as those associated with supercontinent formation, can lower sea levels by reducing the area of the continents. When continents collide, mountains and plateaus are uplifted, and the amount of material taken from below sea level to higher elevations no longer displaces seawater, causing sea levels to drop. The contemporaneous India-Asia collision has caused sea levels to drop by 33 feet (10 m). Times when supercontinents amalgamate are associated with times when seas drop to low levels.

Other factors, such as midplate volcanism, can also change sea levels. The Hawaiian Islands are hot-spot-style midplate volcanoes that have been erupted onto the seafloor, displacing an amount of water equal to their volume. Although this effect is not large at present, at some periods in Earth's history there were many more hot spots (such as in the Cretaceous), and the effect may have been larger.

The effects of the supercontinent cycle on sea level may be summarized as follows. Continent assembly favors regression, whereas continental fragmentation and dispersal favors transgression. Regressions followed formation of the supercontinents of Rodinia and Pangaea, whereas transgressions followed the fragmentation of Rodinia, and the Jurassic-Cretaceous breakup of Pangaea.

Supercontinent Cycle

Movement of the tectonic plates on Earth causes the semi-regular grouping of the planet's landmasses into a single or several large continents that remain stable for a long period of time, then disperse, and eventually come back together as new amalgamated landmasses with a different distribution. This cycle is known as the supercontinent cycle. At several times in Earth history, the continents have joined together to form one large supercontinent, with the last supercontinent Pangaea (meaning all land) breaking up approximately 160 million years ago. This process of supercontinent formation and dispersal and re-amalgamation seems to be grossly cyclic, perhaps reflecting mantle convection patterns, but also influencing sea level, climate, and biological evolution.

The basic idea of the supercontinent cycle is that continents drift about on the surface until they all collide, stay together, and come to rest relative to the mantle. The continents are only half as efficient at conducting heat as oceans, so after the continents are joined together, heat accumulates at their base, causing doming and breakup of the continent. For small continents, heat can flow sideways and not heat up the base of the plate, but for large continents the lateral distance is too great for the heat to be transported sideways. The heat rising from within the Earth therefore breaks up the supercontinent after a heating period of several tens or hundreds of millions of years. The heat then flows away and is transferred to the ocean and atmosphere system, and continents move away until they come back together forming a new supercontinent.

The supercontinent cycle has many effects that greatly affect other Earth systems. First, the breakup of continents causes sudden bursts of heat release, associated with periods of increased, intense magmatism. It also explains some of the large-scale sea-level changes, episodes of rapid and widespread orogeny, episodes of glaciation, and many of the changes in life on Earth.

Sea level has changed by hundreds of meters above and below current levels at many times in Earth history. In fact, sea level is constantly changing in response to a number of different variables, many of them related to plate tectonics. The diversity of fauna on the globe is closely related to sea levels, with greater diversity during sea-level high stands, and lower diversity during sea-level lows. For instance, sea level was 1,970 feet (600 m) higher than now during the ordovician Period, and the sea level high stand was associated with a biotic explosion. Sea levels reached a low stand at the end of the Permian Period, and this low was associated with a great mass extinction. Sea levels were high again in the Cretaceous.

Earthquake

An earthquake is the result of a sudden release of stored energy in the Earth's crust that creates seismic waves. At the Earth's surface, earthquakes may manifest themselves by a shaking or displacement of the ground. Sometimes, they cause tsunamis, which may lead to loss of life and destruction of property. An earthquake is caused by tectonic plates getting stuck and putting a strain on the ground. The strain becomes so great that rocks give way by breaking and sliding along fault planes. Earthquakes are measured with a seismometer, commonly known as a seismograph.

Earthquakes may occur naturally or as a result of human activities. Smaller earthquakes can also be caused by volcanic activity, landslides, mine blasts, and nuclear tests. In its most generic sense, the word earthquake is used to describe any seismic event—whether a natural phenomenon or an event caused by humans—that generates seismic waves.

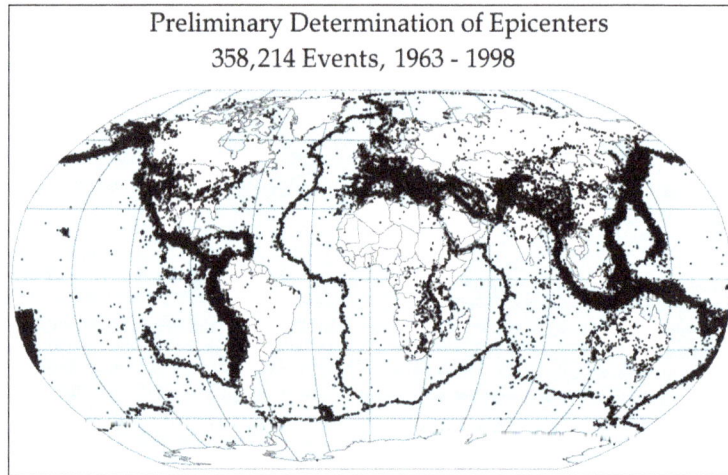

Global earthquake epicenters.

An earthquake's point of initial ground rupture is called its focus or hypocenter. The term epicenter means the point at ground level directly above this.

Naturally Occurring Earthquakes

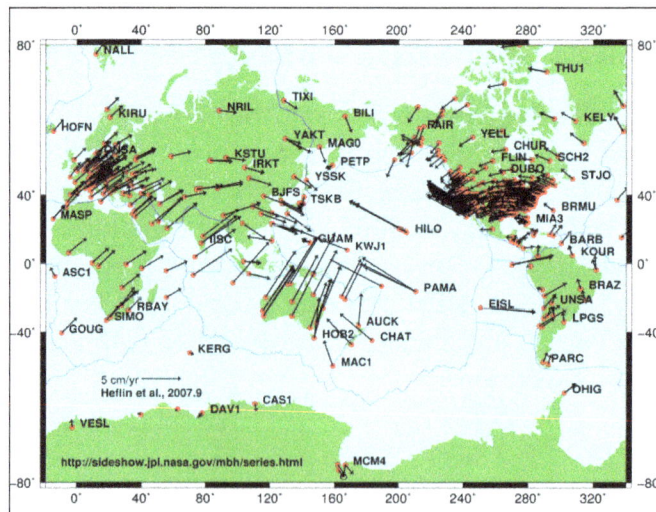

Global plate tectonic movement.

Most naturally occurring earthquakes are related to the tectonic nature of the Earth. Such earthquakes are called tectonic earthquakes. The Earth's lithosphere is a patchwork of plates in slow but constant motion caused by the release to space of the heat in the Earth's mantle and core. The heat causes the rock in the Earth to become flow on geological timescales, so that the plates move, slowly but surely. Plate boundaries lock as the plates move past each other, creating frictional stress. When the frictional stress exceeds a critical value, called local strength, a sudden failure occurs. The boundary of tectonic plates along which failure occurs is called the fault plane. When the failure at the fault plane results in a violent displacement of the Earth's crust, the elastic strain energy is released and seismic waves are radiated, thus causing an earthquake. This process of strain, stress, and failure is referred to as the Elastic-rebound theory. It is estimated that only 10 percent or less of an earthquake's total energy is radiated as seismic energy. Most of the earthquake's energy is used to power the earthquake fracture growth and is converted into heat, or is released to friction. Therefore, earthquakes lower the Earth's available potential energy and raise its temperature, though these changes are negligible.

The majority of tectonic earthquakes originate at depths not exceeding tens of kilometers. In subduction zones, where older and colder oceanic crust descends beneath another tectonic plate, Deep focus earthquakes may occur at much greater depths (up to seven hundred kilometers). These seismically active areas of subduction are known as Wadati-Benioff zones. These are earthquakes that occur at a depth at which the subducted lithosphere should no longer be brittle, due to the high temperature and pressure. A possible mechanism for the generation of deep focus earthquakes is faulting caused by olivine undergoing a phase transition into a spinel structure.

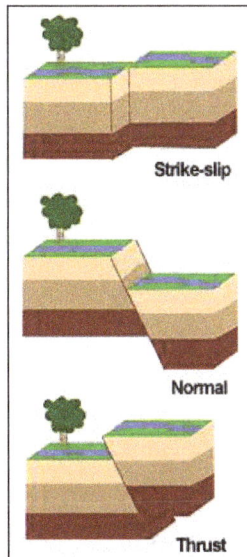

Fault types.

Earthquakes may also occur in volcanic regions and are caused there both by tectonic faults and by the movement of magma in volcanoes. Such earthquakes can be an early warning of volcanic eruptions.

A recently proposed theory suggests that some earthquakes may occur in a sort of earthquake storm, where one earthquake will trigger a series of earthquakes each triggered by the previous shifts on the fault lines, similar to aftershocks, but occurring years later, and with some of the later earthquakes as damaging as the early ones. Such a pattern was observed in the sequence of about a dozen earthquakes that struck the North Anatolian Fault in Turkey in the twentieth century, the half dozen large earthquakes in New Madrid in 1811-1812, and has been inferred for older anomalous clusters of large earthquakes in the Middle East and in the Mojave Desert.

Size and Frequency of Occurrence

Small earthquakes occur nearly constantly around the world in places like California and Alaska in the United States, as well as in Chile, Indonesia, Iran, the Azores in Portugal, New Zealand, Greece, and Japan. Large earthquakes occur less frequently, the relationship being exponential; for example, roughly ten times as many earthquakes larger than magnitude 4 occur in a particular time period than earthquakes larger than magnitude 5. In the (low seismicity) United Kingdom, for example, it has been calculated that the average recurrences are:

- An earthquake of 3.7 or larger every year.

- An earthquake of 4.7 or larger every 10 years.

- An earthquake of 5.6 or larger every 100 years.

The number of seismic stations has increased from about 350 in 1931 to many thousands today. As a result, many more earthquakes are reported than in the past because of the vast improvement in instrumentation (not because the number of earthquakes has increased). The USGS estimates that, since 1900, there have been an average of 18 major earthquakes (magnitude 7.0-7.9) and one great earthquake (magnitude 8.0 or greater) per year, and that this average has been relatively stable. In fact, in recent years, the number of major earthquakes per year has actually decreased, although this is likely a statistical fluctuation. More detailed statistics on the size and frequency of earthquakes is available from the USGS.

Most of the world's earthquakes (90 percent, and 81 percent of the largest) take place in the 40,000-km-long, horseshoe-shaped zone called the circum-Pacific seismic belt, also known as the Pacific Ring of Fire, which for the most part bounds the Pacific Plate. Massive earthquakes tend to occur along other plate boundaries, too, such as along the Himalayan Mountains.

Effects/Impacts of Earthquakes

There are many effects of earthquakes including, but not limited to the following:

Shaking and Ground Rupture

Chūetsu earthquake.

Shaking and ground rupture are the main effects created by earthquakes, principally resulting in more or less severe damage to buildings or other rigid structures. The severity of the local effects depends on the complex combination of the earthquake magnitude, the distance from epicenter, and the local geological and geomorphological conditions, which may amplify or reduce wave propagation. The ground-shaking is measured by ground acceleration.

Smoldering after the 1906 San Francisco earthquake.

Man walking around in Ruins after Tsunami.

Specific local geological, geomorphological, and geo-structural features can induce high levels of shaking on the ground surface even from low-intensity earthquakes. This effect is called site or local amplification. It is principally due to the transfer of the seismic motion from hard deep soils to soft superficial soils and to effects of seismic energy focalization owing to typical geometrical setting of the deposits.

Landslides and Avalanches

Earthquakes can cause landslides and avalanches, which may cause damage in hilly and mountainous areas.

Fires

Following an earthquake, fires can be generated by break of the electrical power or gas lines.

Soil Liquefaction

Soil liquefaction occurs when, because of the shaking, water-saturated granular material temporarily loses their strength and transforms from a solid to a liquid. Soil liquefaction may cause rigid structures, as buildings or bridges, to tilt or sink into the liquefied deposits.

Human Impacts

Earthquakes may result in disease, lack of basic necessities, loss of life, higher insurance premiums, general property damage, road and bridge damage, and collapse of buildings or destabilization of the base of buildings which may lead to collapse in future earthquakes.

Volcanology

Volcanology, also spelled vulcanology is the discipline of the geologic sciences that is concerned with all aspects of volcanic phenomena.

Volcanology deals with the formation, distribution, and classification of volcanoes as well as with their structure and the kinds of materials ejected during an eruption (such as pyroclastic flows, lava, dust, ash, and volcanic gases). It also involves research on the relationships between volcanic eruptions and other large-scale geologic processes such as plate tectonics, mountain building, and earthquakes. One of the chief objectives of this research is determining the nature and causes of volcanic eruptions for the purpose of forecasting their occurrence. Another practical concern of volcanology is securing data that may aid in locating commercially valuable deposits of ores, particularly those of certain sulfide minerals.

Volcano

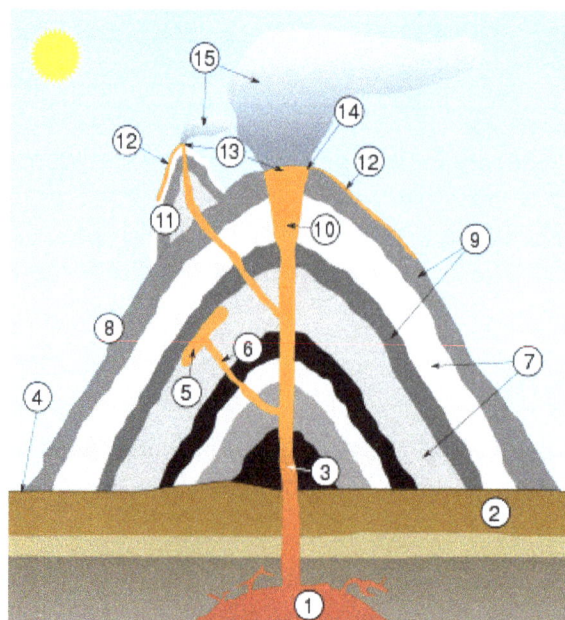

Volcano (Figure):	
1. Large magma chamber 2. Bedrock 3. Conduit (pipe) 4. Base 5. Sill 6. Branch pipe 7. Layers of ash emitted by the volcano 8. Flank	9. Layers of lava emitted by the volcano 10. Throat 11. Parasitic cone 12. Lava flow 13. Vent 14. Crater 15. Ash cloud

A volcano is an opening, or rupture, in the Earth's surface or crust, which allows hot, molten rock, ash, and gases to escape from deep below the surface. Volcanic activity involving the extrusion of rock tends to form mountains or features like mountains over a period of time.

Volcanoes are generally found where tectonic plates pull apart or come together. A mid-oceanic ridge, like the Mid-Atlantic Ridge, has examples of volcanoes caused by "divergent tectonic plates"—that is, plates pulling apart. The Pacific Ring of Fire has examples of volcanoes caused by "convergent tectonic plates"—that is, plates coming together. By contrast, volcanoes are usually not created where two tectonic plates slide past each other. Volcanoes can also form where the Earth's crust stretches and grows thin, called "non-hotspot intraplate volcanism"—examples include the African Rift Valley, the European Rhine Graben with its Eifel volcanoes, the Wells Gray-Clearwater Volcanic Field, and the Rio Grande Rift in North America.

Finally, volcanoes can be caused by "mantle plumes," so-called "hotspots." These hotspots can occur far from plate boundaries, such as the Hawaiian Islands. Interestingly, hotspot volcanoes are also found elsewhere in the Solar System, especially on rocky planets and moons.

Locations

Divergent Plate Boundaries

At the mid-oceanic ridges, two tectonic plates diverge from one another. New oceanic crust is being formed by hot molten rock slowly cooling down and solidifying. In these places, the crust is very thin due to the pull of the tectonic plates. The release of pressure due to the thinning of the crust leads to adiabatic expansion, and the partial melting of the mantle. This melt causes the volcanism and make the new oceanic crust. The main part of the mid-oceanic ridges is at the bottom of the ocean, and most volcanic activity is submarine. Black smokers are a typical example of this kind of volcanic activity. Where the mid-oceanic ridge comes above sea-level, volcanoes like the Hekla on Iceland are formed. Divergent plate boundaries create new seafloor and volcanic islands.

Convergent Plate Boundaries

"Subduction zones," as they are called, are places where two plates, usually an oceanic plate and a continental plate, collide. In this case, the oceanic plate subducts (submerges) under the continental plate forming a deep ocean trench just offshore. The crust is then melted by the heat from the mantle and becomes magma. This is due to the water content lowering the melting temperature. The magma created here tends to be very viscous due to its high silica content, so often does not reach the surface and cools at depth. When it does reach the surface, a volcano is formed. Typical examples for this kind of volcano are the volcanoes in the Pacific Ring of Fire, Mount Etna.

Hotspots

Hotspots are not located on the ridges of tectonic plates, but on top of mantle plumes, where the convection of Earth's mantle creates a column of hot material that rises until it reaches the crust, which tends to be thinner than in other areas of the Earth. The temperature of the plume causes the crust to melt and form pipes, which can vent magma. Because the tectonic plates move whereas the mantle plume remains in the same place, each volcano becomes dormant after a while and a new volcano is then formed as the plate shifts over the hotspot. The Hawaiian Islands are thought to be formed in such a manner, as well as the Snake River Plain, with the Yellowstone Caldera being the current part of the North American plate over the hotspot.

Volcanic Features

The most common perception of a volcano is of a conical mountain, spewing lava and poisonous gases from a crater in its top. This describes just one of many types of volcano, and the features of volcanoes are much more complicated. The structure and behavior of volcanoes depends on a number of factors. Some volcanoes have rugged peaks formed by lava domes rather than a summit crater, whereas others present landscape features such as massive plateaus. Vents that issue volcanic material (lava, which is what magma is called once it has broken the surface, and ash) and gases (mainly steam and magmatic gases) can be located anywhere on the landform.

Other types of volcanoes include cryovolcanos (or ice volcanoes), particularly on some moons of Jupiter, Saturn and Neptune; and mud volcanoes, which are formations often not associated with known magmatic activity. Active mud volcanoes tend to involve temperatures much lower than those of igneous volcanoes, except when a mud volcano is actually a vent of an igneous volcano.

Shield Volcanoes

Hawaii and Iceland are examples of places where volcanoes extrude huge quantities of basaltic lava that gradually build a wide mountain with a shield-like profile. Their lava flows are generally very hot and very fluid, contributing to long flows. The largest lava shield on Earth, Mauna Loa, rises over 9,000 m from the ocean floor, is 120 km in diameter and forms part of the Big Island of Hawaii, along with other shield volcanoes such as Mauna Kea and Kīlauea. Olympus Mons is the largest shield volcano on Mars, and is the tallest known mountain in the solar system. Smaller versions of shield volcanoes include *lava cones,* and *lava mounds.*

Quiet eruptions spread out basaltic lava in flat layers. The buildup of these layers form a broad volcano with gently sloping sides called a shield volcano. Examples of shield volcanoes are the Hawaiian Islands.

Toes of a pāhoehoe advance across a road in Kalapana on the east
rift zone of Kīlauea Volcano in Hawaii.

Cinder Cones

Volcanic cones or *cinder cones* result from eruptions that throw out mostly small pieces of scoria and pyroclastics (both resemble cinders, hence the name of this volcano type) that build up around the vent. These can be relatively short-lived eruptions that produce a cone-shaped hill perhaps 30 to 400 m high. Most cinder cones erupt only once. Cinder cones may form as flank vents on larger volcanoes, or occur on their own. Parícutin in Mexico and Sunset Crater in Arizona are examples of cinder cones.

Stratovolcanoes

Stratovolcanoes are tall conical mountains composed of lava flows and other ejecta in alternate layers, the strata that give rise to the name. Stratovolcanoes are also known as composite volcanoes. Classic examples include Mt. Fuji in Japan, Mount Mayon in the Philippines, and Mount Vesuvius and Stromboli in Italy.

Aa is a term of Polynesian origin, pronounced Ah-ah,
for rough, jagged, spiny lavaflow.

Super Volcanoes

A *supervolcano* is the popular term for a large volcano that usually has a large caldera and can potentially produce devastation on an enormous, sometimes continental, scale. Such eruptions would be able to cause severe cooling of global temperatures for many years afterwards because of the huge volumes of sulfur and ash erupted. They can be the most dangerous type of volcano. Examples include Yellowstone Caldera in Yellowstone National Park, Lake Taupo in New Zealand and Lake Toba in Sumatra, Indonesia. Supervolcanoes are hard to identify centuries later, given the enormous areas they cover. Large igneous provinces are also considered supervolcanoes because of the vast amount of basalt lava erupted.

Submarine Volcanoes

Submarine volcanoes are common features on the ocean floor. Some are active and, in shallow water, disclose their presence by blasting steam and rocky debris high above the surface of the sea. Many others lie at such great depths that the tremendous weight of the water above them prevents the explosive release of steam and gases, although they can be detected by hydrophones and discoloration of water because of volcanic gases. Even large submarine eruptions may not disturb the ocean surface. Because of the rapid cooling effect of water as compared to air, and increased buoyancy, submarine volcanoes often form rather steep pillars over their volcanic vents as compared to above-surface volcanoes. In due time, they may break the ocean surface as new islands. Pillow lava is a common eruptive product of submarine volcanoes.

Pillow lava (NOAA).

Subglacial Volcanoes

Herðubreið, one of the tuyas in Iceland.

Subglacial *volcanoes* develop underneath icecaps. They are made up of flat lava flows atop extensive pillow lavas and palagonite. When the icecap melts, the lavas on the top collapse leaving a flat-topped mountain. Then, the pillow lavas also collapse, giving an angle of 37.5 degrees. These volcanoes are also called Table Mountains, tuyas or (uncommonly) mobergs. Very good examples of this type of volcano can be seen in Iceland; however, there are also tuyas in British Columbia. The origin of the term comes from Tuya Butte, which is one of the several tuyas in the area of the Tuya River and Tuya Range in northern British Columbia. Tuya Butte was the first such landform analyzed and so its name has entered the geological literature for this kind of volcanic formation. The Tuya Mountains Provincial Park was recently established to protect this unusual landscape, which lies north of Tuya Lake and south of the Jennings River near the boundary with the Yukon Territory.

Erupted Material

Lava Composition

Another way of classifying volcanoes is by the *composition of material erupted* (lava), since this affects the shape of the volcano. Lava can be broadly classified into 4 different compositions:

- If the erupted magma contains a high percentage (more than 63 percent) of silica, the lava is called felsic.

    ○ Felsic lavas (or rhyolites) tend to be highly viscous (not very fluid) and are erupted as domes or short, stubby flows. Viscous lavas tend to form stratovolcanoes or lava domes. Lassen Peak in California is an example of a volcano formed from felsic lava and is actually a large lava dome.

    ○ Because siliceous magmas are so viscous, they tend to trap volatiles (gases) that are present, which cause the magma to erupt catastrophically, eventually forming strato-volcanoes. Pyroclastic flows (ignimbrites) are highly hazardous products of such volcanoes, since they are composed of molten volcanic ash too heavy to go up into the atmosphere, so they hug the volcano's slopes and travel far from their vents during large eruptions. Temperatures as high as 1,200 °C are known to occur in pyroclastic flows, which will incinerate everything flammable in their path and thick layers of hot pyroclastic flow deposits can be laid down, often up to many meters thick. Alaska's Valley of Ten Thousand Smokes, formed by the eruption of Novarupta near Katmai in 1912, is an example of a thick pyroclastic flow or ignimbrite deposit. Volcanic ash that is light enough to be erupted high into the Earth's atmosphere may travel many kilometres before it falls back to ground as a tuff.

- If the erupted magma contains 52-63 percent silica, the lava is of *intermediate* composition.

    ○ These "andesitic" volcanoes generally only occur above subduction zones (for example, Mount Merapi in Indonesia).

- If the erupted magma contains between 45 and 52 percent silica, the lava is called mafic (because it contains higher percentages of magnesium (Mg) and iron (Fe)) or basaltic. These lavas are usually much less viscous than rhyolitic lavas, depending on their eruption temperature; they also tend to be hotter than felsic lavas. Mafic lavas occur in a wide range of settings:

- At mid-ocean ridges, where two oceanic plates are pulling apart, basaltic lava erupts as pillowsto fill the gap;

- Shield volcanoes (e.g. the Hawaiian Islands, including Mauna Loa and Kilauea), on both oceanic and continental crust;

- As continental flood basalts.

- Some erupted magmas contain up to 45 percent silica and produce lava called ultramafic. Ultramafic flows, also known as komatiites, are very rare; indeed, very few have been erupted at the Earth's surface since the Proterozoic, when the planet's heat flow was higher. They are (or were) the hottest lavas, and probably more fluid than common mafic lavas.

Lava Texture

Two types of lava are named according to the surface texture: 'A'a and pāhoehoe, both words having Hawaiian origins. 'A'a is characterized by a rough, clinkery surface and is what most viscous and hot lava flows look like. However, even basaltic or mafic flows can be erupted as 'a'a flows, particularly if the eruption rate is high and the slope is steep. Pahoehoe is characterized by its smooth and often ropey or wrinkly surface and is generally formed from more fluid lava flows. Usually, only mafic flows will erupt as pāhoehoe, since they often erupt at higher temperatures or have the proper chemical make-up to allow them to flow at a higher fluidity.

Volcanic Activity

A volcanic fissure and lava channel.

A popular way of classifying magmatic volcanoes goes by their frequency of eruption, with those that erupt regularly called active, those that have erupted in historical times but are now quiet called dormant, and those that have not erupted in historical times called extinct. However, these popular classifications—extinct in particular—are practically meaningless to scientists. They use classifications which refer to a particular volcano's formative and eruptive processes and resulting shapes.

Mount St. Helens in 1980, shortly after the eruption on May 18.

There is no real consensus among volcanologists on how to define an "active" volcano. The lifespan of a volcano can vary from months to several million years, making such a distinction sometimes meaningless when compared to the lifespans of humans or even civilizations. For example, many of Earth's volcanoes have erupted dozens of times in the past few thousand years but are not currently showing signs of eruption. Given the long lifespan of such volcanoes, they are very active. By our lifespans, however, they are not. Complicating the definition are volcanoes that become restless (producing earthquakes, venting gases, or other non-eruptive activities) but do not actually erupt.

Scientists usually consider a volcano active if it is currently erupting or showing signs of unrest, such as unusual earthquake activity or significant new gas emissions. Many scientists also consider a volcano active if it has erupted in historic time. It is important to note that the span of recorded history differs from region to region; in the Mediterranean, recorded history reaches back more than 3,000 years but in the Pacific Northwest of the United States, it reaches back less than 300 years, and in Hawaii, little more than 200 years. The Smithsonian Global Volcanism Program's definition of 'active' is having erupted within the last 10,000 years.

Dormant volcanoes are those that are not currently active (as defined above), but could become restless or erupt again. Confusion however, can arise because many volcanoes which scientists consider to be *active* are referred to as *dormant* by laypersons or in the media.

Extinct volcanoes are those that scientists consider unlikely to erupt again. Whether a volcano is truly extinct is often difficult to determine. Since "supervolcano" calderas can have eruptive lifespans sometimes measured in millions of years, a caldera that has not produced an eruption in tens of thousands of years is likely to be considered dormant instead of extinct.

For example, the Yellowstone Caldera in Yellowstone National Park is at least two million years old and hasn't erupted violently for approximately 640,000 years, although there has been some minor activity relatively recently, with hydrothermal eruptions less than 10,000 years ago and lava flows about 70,000 years ago. For this reason, scientists do not consider the Yellowstone Caldera extinct. In fact, because the caldera has frequent earthquakes, a very active geothermal system (i.e., the entirety of the geothermal activity found in Yellowstone National Park), and rapid rates of ground uplift, many scientists consider it to be an active volcano.

Notable Volcanoes

On Earth

The Decade Volcanoes are 17 volcanoes identified by the International Association of Volcanology and Chemistry of the Earth's Interior (IAVCEI) as being worthy of particular study in light of their history of large, destructive eruptions and proximity to populated areas. They are named Decade Volcanoes because the project was initiated as part of the United Nations-sponsored International Decade for Natural Disaster Reduction. The 17 current Decade Volcanoes are:

• Avachinsky-Koryaksky (grouped together), Kamchatka, Russia	• Sakurajima, Kagoshima Prefecture, Japan
• Nevado de Colima, Jalisco and Colima, Mexico	• Santa Maria/Santiaguito, Guatemala
• Mount Etna, Sicily, Italy	• Santorini, Cyclades, Greece
• Galeras, Nariño, Colombia	• Taal Volcano, Luzon, Philippines
• Mauna Loa, Hawaii, USA	• Teide, Canary Islands, Spain
• Mount Merapi, Central Java, Indonesia	• Ulawun, New Britain, Papua New Guinea
• Mount Nyiragongo, Democratic Republic of the Congo	• Mount Unzen, Nagasaki Prefecture, Japan
• Mount Rainier, Washington, USA	• Vesuvius, Naples, Italy

Elsewhere in the Solar System

The Earth's Moon has no large volcanoes and no current volcanic activity, although recent evidence suggests it may still possess a partially molten core. However, the Moon does have many volcanic features such as maria (the darker patches seen on the moon), rilles and domes.

The planet Venus has a surface that is 90 percent basalt, indicating that volcanism played a major role in shaping its surface. The planet may have had a major global resurfacing event about 500 million years ago from what scientists can tell from the density of impact craters on the surface. Lava flows are widespread and forms of volcanism not present on Earth occur as well. Changes in the planet's atmosphere and observations of lightning have been attributed to ongoing volcanic eruptions, although there is no confirmation of whether or not Venus is still volcanically active.

Olympus Mons (Latin, "Mount Olympus") is the tallest known mountain in
our solar system, located on the planet Mars.

There are several extinct volcanoes on Mars, four of which are vast shield volcanoes far bigger than any on Earth. They include Arsia Mons, Ascraeus Mons, Hecates Tholus, Olympus Mons, and Pavonis Mons. These volcanoes have been extinct for many millions of years, but the European *Mars Express* spacecraft has found evidence that volcanic activity may have occurred on Mars in the recent past as well.

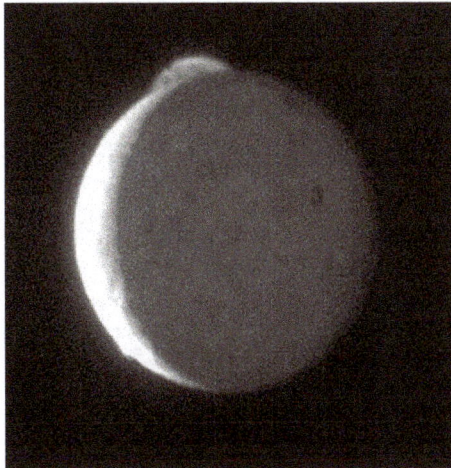

The Tvashtar volcano erupts a plume 330 km (205 mi)
above the surface of Jupiter's moon Io.

Jupiter's moon Io is the most volcanically active object in the solar system because of tidal interaction with Jupiter. It is covered with volcanoes that erupt sulfur, sulfur dioxide and silicate rock, and as a result, Io is constantly being resurfaced. Its lavas are the hottest known anywhere in the solar system, with temperatures exceeding 1,800 K (1,500 °C). In February 2001, the largest recorded volcanic eruptions in the solar system occurred on Io. Europa, the smallest of Jupiter's Galilean moons, also appears to have an active volcanic system, except that its volcanic activity is entirely in the form of water, which freezes into ice on the frigid surface. This process is known as cryovolcanism, and is apparently most common on the moons of the outer planets of the solar system.

In 1989 the Voyager 2 spacecraft observed cryovolcanos (ice volcanoes) on Triton, a moon of Neptune, and in 2005 the Cassini-Huygens probe photographed fountains of frozen particles erupting from Enceladus, a moon of Saturn. The ejecta may be composed of water, liquid nitrogen, dust, or methanecompounds. Cassini-Huygens also found evidence of a methane-spewing cryovolcano on the Saturnianmoon Titan, which is believed to be a significant source of the methane found in its atmosphere. It is theorized that cryovolcanism may also be present on the Kuiper Belt Object Quaoar.

Effects of Volcanoes

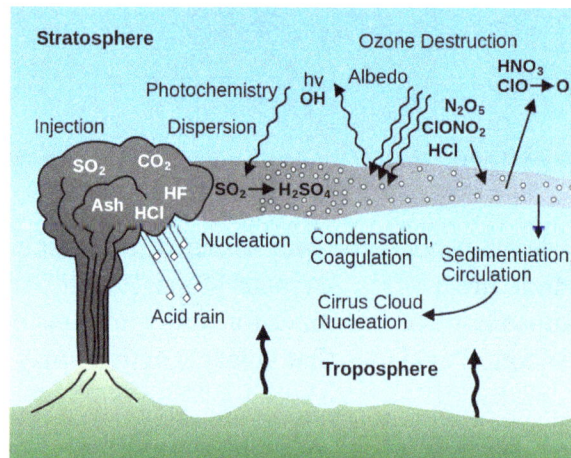

Volcanic "injection".

There are many different kinds of volcanic activity and eruptions: phreatic eruptions (steam-generated eruptions), explosive eruption of high-silica lava (e.g., rhyolite), effusive eruption of low-silica lava (e.g., basalt), pyroclastic flows, lahars (debris flow) and carbon dioxideemission. All of these activities can pose a hazard to humans. Earthquakes, hot springs, fumaroles, mud pots and geysers often accompany volcanic activity.

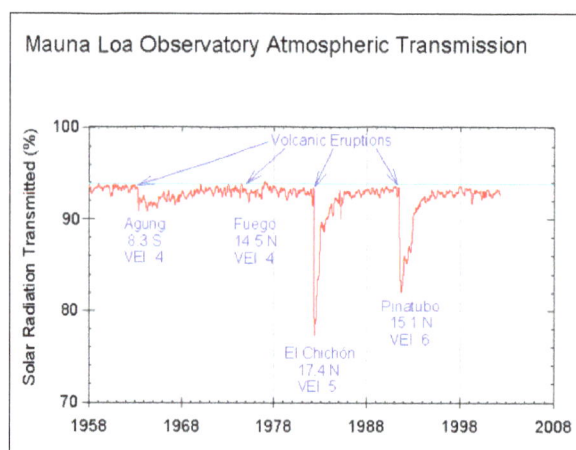

Solar radiation reduction from volcanic eruptions.

The concentrations of different volcanic gases can vary considerably from one volcano to the next. Water vapor is typically the most abundant volcanic gas, followed by carbon dioxide and sulphur dioxide. Other principal volcanic gases include hydrogen sulphide, hydrogen chloride, and hydrogen fluoride. A large number of minor and trace gases are also found in volcanic emissions, for example hydrogen, carbon monoxide, halocarbons, organic compounds, and volatile metal chlorides.

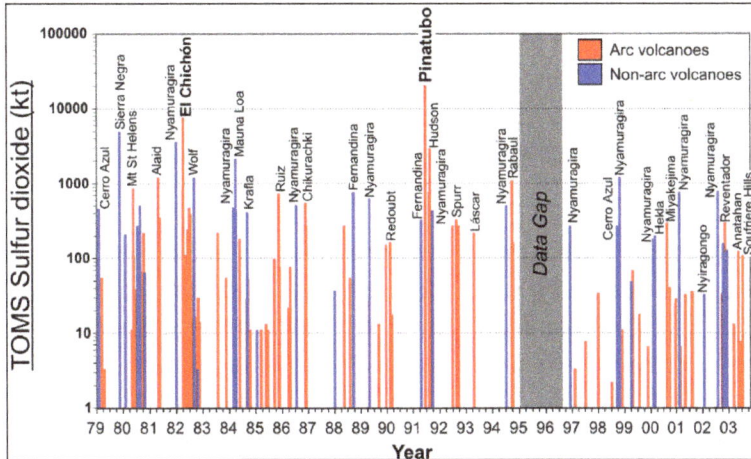

Sulfur dioxide emissions by volcanoes.

Large, explosive volcanic eruptions inject water vapor (H_2O), carbon dioxide (CO_2), sulfur dioxide (SO_2), hydrogen chloride (HCl), hydrogen fluoride (HF) and ash (pulverized rock and pumice) into the stratosphere to heights of 10-20 miles above the Earth's surface. The most significant impacts from these injections come from the conversion of sulphur dioxide to sulphuric acid (H_2SO_4), which condenses rapidly in the stratosphere to form fine sulfate aerosols. The aerosols increase the Earth's albedo—its reflection of radiation from the Sun back into space - and thus cool the Earth's lower atmosphere or troposphere; however, they also absorb heat radiated up from the Earth, thereby warming the stratosphere.

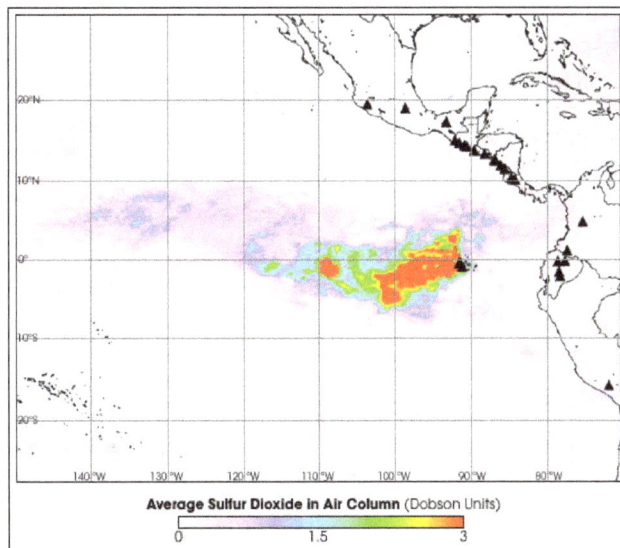

Average concentration of sulfur dioxide over the Sierra Negra Volcano
(Galapagos Islands) from October 23-November 1, 2005.

Several eruptions during the past century have caused a decline in the average temperature at the Earth's surface of up to half a degree (Fahrenheit scale) for periods of one to three years. The sulphate aerosols also promote complex chemical reactions on their surfaces that alter chlorine and nitrogen chemical species in the stratosphere. This effect, together with increased stratospheric chlorine levels from chlorofluorocarbon pollution, generates chlorine monoxide (ClO), which destroys ozone (O_3). As the aerosols grow and coagulate, they settle down into the upper troposphere where they serve as nuclei for cirrus clouds and further modify the Earth's radiation balance. Most of the hydrogen chloride (HCl) and hydrogen fluoride (HF) are dissolved in water droplets in the eruption cloud and quickly fall to the ground as acid rain. The injected ash also falls rapidly from the stratosphere; most of it is removed within several days to a few weeks. Finally, explosive volcanic eruptions release the greenhouse gas carbon dioxide and thus provide a deep source of carbon for biogeochemical cycles.

Gas emissions from volcanoes are a natural contributor to acid rain. Volcanic activity releases about 130 to 230 teragrams (145 million to 255 million short tons) of carbon dioxide each year. Volcanic eruptions may inject aerosols into the Earth's atmosphere. Large injections may cause visual effects such as unusually colorful sunsets and affect global climate mainly by cooling it. Volcanic eruptions also provide the benefit of adding nutrients to soil through the weathering process of volcanic rocks. These fertile soils assist the growth of plants and various crops. Volcanic eruptions can also create new islands, as the magma cools and solidifies upon contact with the water.

References

- Plate-tectonics: newworldencyclopedia.org, Retrieved 9 February, 2019

- Plate-tectonics, science: britannica.com, Retrieved 28 August, 2019

- Tectonic-landforms: geography.name, Retrieved 22 February, 2019

- Supercontinents-and-climate, plate-tectonics: climate-policy-watcher.org, Retrieved 20 July, 2019

- Plate-tectonics-supercontinent-cycles-and-sea-level, plate-tectonics: climate-policy-watcher.org, Retrieved 2 May, 2019

- Earthquake: newworldencyclopedia.org, Retrieved 1 March, 2019

- Volcanology, science: britannica.com, Retrieved 25 January, 2019

- Volcano: newworldencyclopedia.org, Retrieved 5 June, 2019

Chapter 4

Mineralogy and Petrology

The branch of geology which studies the crystal structure, chemistry and physical characteristics of minerals and mineralized artifacts is termed as mineralogy. Petrology refers to the branch geology which deals with the study of rocks and the conditions under which they form. The topics elaborated in this chapter will help in gaining a better perspective about the diverse aspects of mineralogy and petrology.

Mineralogy

Mineralogy is an Earth science focused around the chemistry, crystal structure, and physical (including optical) properties of minerals. Specific studies within mineralogy include the processes of mineral origin and formation, classification of minerals, their geographical distribution, as well as their utilization.

Physical Mineralogy

Physical mineralogy is the specific focus on physical attributes of minerals. Description of physical attributes is the simplest way to identify, classify, and categorize minerals, and they include:

- Crystal structure
- Crystal habit
- Twinning
- Cleavage
- Luster
- Color
- Streak
- Hardness
- Specific gravity

Chemical Mineralogy

Chemical mineralogy focuses on the chemical composition of minerals in order to identify, classify, and categorize them, as well as a means to find beneficial uses from them. There are a few minerals which are classified as whole elements, including sulfur, copper, silver, and gold, yet the vast majority of minerals are comprised of chemical compounds, some more complex than others. In terms of major chemical divisions of minerals, most are placed within the isomorphous groups, which are based on analogous chemical composition and similar crystal forms. A good example of isomorphism classification would be the calcite group, containing the minerals calcite, magnesite, siderite, rhodochrosite, and smithsonite.

Biomineralogy

Biomineralogy is a cross-over field between mineralogy, paleontology and biology. It is the study of how plants and animals stabilize minerals under biological control, and the sequencing of mineral replacement of those minerals after deposition. It uses techniques from chemical mineralogy, especially isotopic studies, to determine such things as growth forms in living plants and animals as well as things like the original mineral content of fossils.

Optical Mineralogy

Optical mineralogy is a specific focus of mineralogy that applies sources of light as a means to identify and classify minerals. All minerals which are not part of the cubic system are double refracting, where ordinary light passing through them is broken up into two plane polarized rays that travel at different velocities and refracted at different angles. Mineral substances belonging to the cubic system pertain only one index of refraction. Hexagonal and tetragonal mineral substances have two indices, while orthorhombic, monoclinic, and triclinic substances have three indices of refraction. With opaque ore minerals, reflected light from a microscope is needed for identification.

Crystal Structure

The use of X-ray to determine the atomic arrangement of minerals is also another way to identify and classify minerals. With minerals pertaining highly complex compositions, the exact formula of the mineral's composition can be easily discerned with knowledge of its structure. The structure of a mineral also offers a precise way of establishing isomorphism. With crystal structure, one may also deduce the correlation between atomic positions and specific physical properties.

Formation and Occurrence

The effects of provided by variables and catalysts such as pressure, temperature, and time allow for the process of the formation of minerals. This process can range from simple processes found in nature, to complex formations that take years or even centuries of time. The origins of certain minerals are certainly obvious, with those such as rock salt and gypsum from evaporating sea water. Various possible methods of formation include:

- Sublimation from volcanic gases;

- Deposition from aqueous solutions and hydrothermal brines;

- Crystallization from an igneous magma or lava;

- Recrystallization due to metamorphic processes and metasomatism;

- Crystallization during diagenesis of sediments;

- Formation by oxidation and weathering of rocks exposed to the atmosphere or soil environment.

Uses

Minerals are essential to various needs within human society, such as minerals used for bettering healthand fitness (such as mineral water or commercially-sold vitamins), essential components of metal products used in various commodities and machinery, essential components to building materials such as limestone, marble, granite, gravel, glass, plaster, cement, plastics, etc. Minerals are also used in fertilizers to enrich the growth of agricultural crops.

Descriptive Mineralogy

Descriptive mineralogy summarizes results of studies performed on mineral substances. It is the scholarly and scientific method of recording the identification, classification, and categorization of minerals, their properties, and their uses. Classifications for descriptive mineralogy follow as such:

- Elements;

- Sulfides;

- Oxides and hydroxides;

- Halides;

- Nitrates, carbonates, and borates;

- Sulfates, chromates, molybdates, and tungstates;

- Phosphates, arsenates, and vanadates;

- Silicates.

Determinative Mineralogy

Determinative mineralogy is the actual scientific process of identifying minerals, through data gathering and conclusion. When new minerals are discovered, a standard procedure of scientific analysis is followed, including measures to identify a mineral's formula, its crystallographic data, its optical data, as well as the general physical attributes determined and listed.

Mineral

Mineral is a naturally occurring homogeneous solid with a definite chemical composition and a highly ordered atomic arrangement; it is usually formed by inorganic processes. There are several thousand known mineral species, about 100 of which constitute the major mineral components of rocks; these are the so-called rock-forming minerals.

A mineral, which by definition must be formed through natural processes, is distinct from the synthetic equivalents produced in the laboratory. Artificial versions of minerals, including emeralds, sapphires, diamonds, and other valuable gemstones, are regularly produced in industrial and research facilities and are often nearly identical to their natural counterparts.

Amethyst: Amethyst, a well-known gemstone, is
a variety of the silica mineral quartz.

By its definition as a homogeneous solid, a mineral is composed of a single solid substance of uniform composition that cannot be physically separated into simpler chemical compounds. Homogeneity is determined relative to the scale on which it is defined. A specimen that appears homogeneous to the unaided eye, for example, may reveal several mineral components under a microscope or upon exposure to X-ray diffraction techniques. Most rocks are composed of several different minerals; e.g., granite consists of feldspar, quartz, mica, and amphibole. In addition, gasesand liquids are excluded by a strict interpretation of the above definition of a mineral. Ice, the solid state of water (H_2O), is considered a mineral, but liquid water is not; liquid mercury, though sometimes found in mercury oredeposits, is not classified as a mineral either. Such substances that resemble minerals in chemistry and occurrence are dubbed mineraloids and are included in the general domain of mineralogy.

Trigonal system: Trigonal (rhombohedral) crystals of quartz.

Since a mineral has a definite composition, it can be expressed by a specific chemical formula. Quartz (silicon dioxide), for instance, is rendered as SiO_2, because the elements silicon (Si) and oxygen (O) are it's only constituentsand they invariably appear in a 1:2 ratio. The chemical make-up of most minerals is not as well defined as that of quartz, which is a pure substance. Siderite, for example, does not always occur as pure iron carbonate ($FeCO_3$); magnesium (Mg), manganese (Mn), and, to a limited extent, calcium (Ca) may sometimes substitute for the iron. Since the amount of the replacement may vary, the composition of siderite is not fixed and ranges between certain limits, although the ratio of the metal cation to the anionicgroup remains fixed at 1:1. Its chemical makeup may be expressed by the general formula (Fe, Mn, Mg, Ca)CO_3, which reflects the variability of the metal content.

Minerals display a highly ordered internal atomic structure that has a regular geometric form. Because of this feature, minerals are classified as crystalline solids. Under favourable conditions, crystalline materials may express their ordered internal framework by a well-developed external form, often referred to as crystal form or morphology. Solids that exhibit no such ordered internal arrangement are termed amorphous. Many amorphous natural solids, such as glass, are categorized as mineraloids.

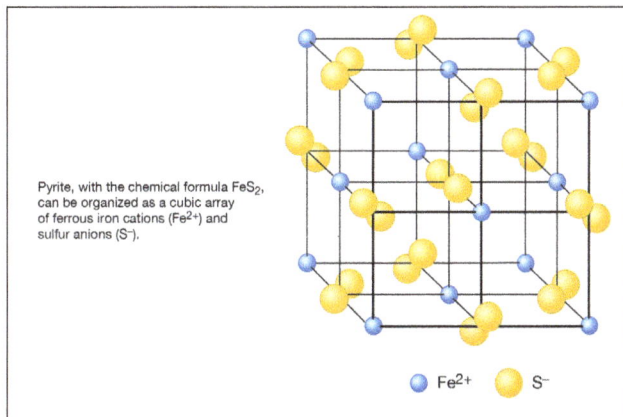

Pyrite, with the chemical formula FeS_2, can be organized as a cubic array of ferrous iron cations (Fe^{2+}) and sulfur anions (S^-).

Pyrite: Schematic representation of the structure of pyrite (FeS_2) as based on a cubic array of ferrous iron cations (Fe^{2+}) and sulfur anions (S^-).

Traditionally, minerals have been described as resulting exclusively from inorganic processes; however, current mineralogic practice often includes as minerals those compounds that are organically produced but satisfy all other mineral requirements. Aragonite ($CaCO_3$) is an example of an inorganically formed mineral that also has an organically produced, yet otherwise identical, counterpart; the shell (and the pearl, if it is present) of an oyster is composed to a large extent of organically formed aragonite. Minerals also are produced by the human body: hydroxylapatite [$Ca_5(PO_4)_3(OH)$] is the chief component of bones and teeth, and calculi are concretions of mineral substances found in the urinary system.

Occurrence and Formation

Minerals form in all geologic environments and thus under a wide range of chemical and physical conditions, such as varying temperature and pressure. The four main categories of mineral formation are: (1) igneous, or magmatic, in which minerals crystallize from a melt, (2) sedimentary, in which minerals are the result of sedimentation, a process whose raw materials are particles from other rocks that have undergone weathering or erosion, (3) metamorphic, in which new minerals form at the expense of earlier ones owing to the effects of changing—usually increasing—temperature or pressure or both on some existing rock type, and (4) hydrothermal, in which minerals are chemically precipitated from hot solutions within Earth. The first three processes generally lead to varieties of rocks in which different mineral grains are closely intergrown in an interlocking fabric. Hydrothermal solutions, and even solutions at very low temperatures (e.g., groundwater), tend to follow fracture zones in rocks that may provide open spaces for the chemical precipitation of minerals from solution. It is from such open spaces, partially filled by minerals deposited from solutions, that most of the spectacular mineral specimens have been collected. If a mineral that is in the process of growth (as a result of precipitation) is allowed to develop in a free space, it will generally exhibit a well-developed crystal form, which adds to a specimen's aesthetic beauty. Similarly, geodes, which are rounded, hollow, or partially hollow bodies commonly found in limestones, may contain well-formed crystals lining the central cavity. Geodes form as a result of mineral deposition from solutions such as groundwater.

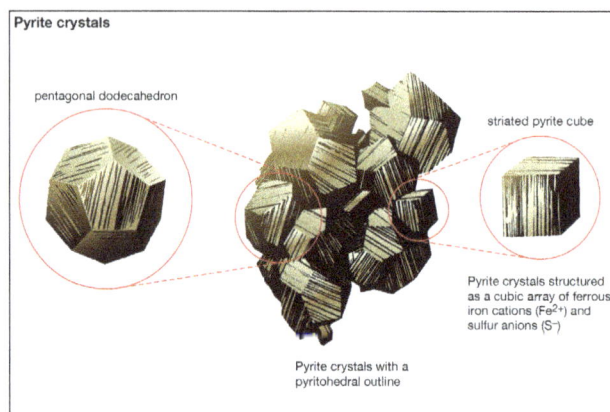

Pyrite crystals: Pyrite crystal shape depends on the amount of iron and sulfur in the sample. Iron-rich pyrite forms cube-shaped crystals, whereas samples that contain increasing amounts of sulfur form duodecahedral-shaped crystals.

The Nature Of Minerals

Morphology

Azurite: Azurite crystals.

Nearly all minerals have the internal ordered arrangement of atoms and ions that is the defining characteristic of crystalline solids. Under favourable conditions, minerals may grow as well-formed crystals, characterized by their smooth plane surfaces and regular geometric forms. Development of this good external shape is largely a fortuitous outcome of growth and does not affect the basic properties of a crystal. Therefore, the term *crystal* is most often used by material scientists to refer to any solid with an ordered internal arrangement, without regard to the presence or absence of external faces.

Symmetry Elements

The external shape, or morphology, of a crystal is perceived as its aesthetic beauty, and its geometry reflects the internal atomic arrangement. The external shape of well-formed crystals expresses the presence or absence of a number of symmetry elements. Such symmetry elements include rotation axes, rotoinversion axes, a centre of symmetry, and mirror planes.

Cueva de los Cristales: Massive selenite (gypsum) crystals from the Cave of Crystals (Cueva de los Cristales), Naica Mine, Chihuahua, Mexico.

A rotation axis is an imaginary line through a crystal around which it may be rotated and repeat itself in appearance one, two, three, four, or six times during a complete rotation. (For example, a sixfold rotation occurs when the crystal repeats itself each 60°—that is, six times in a 360° rotation).

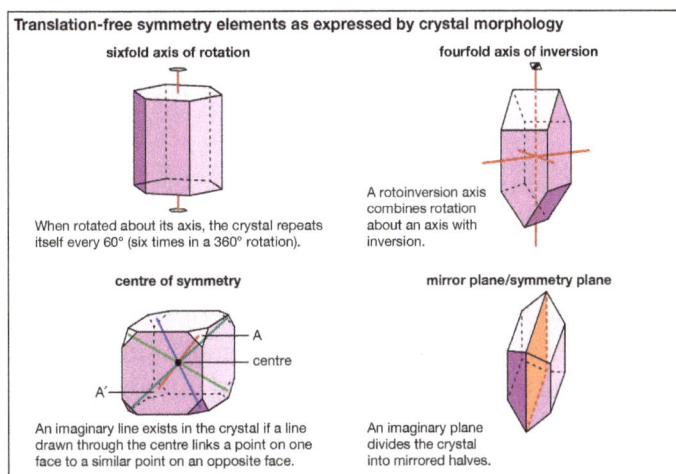

Symmetry elements: Translation-free symmetry elements as expressed by the morphology of crystals showing a sixfold axis of rotation, fourfold axis of inversion, centre of symmetry, and mirror/symmetry plane.

A rotoinversion axis combines rotation about an axis of rotation with inversion. Rotoinversion axes are symbolized as $\bar{1}, \bar{2}, \bar{3}, \bar{4}$, and $\bar{6}$, where $\bar{1}$ is equivalent to a centre of symmetry (or inversion), $\bar{2}$ is equivalent to a mirror plane, and $\bar{3}$ is equivalent to a threefold rotation axis plus a centre of symmetry. When the axis of the crystal is vertical, $\bar{4}$ is characterized by two top faces with identical faces upside down underneath. $\bar{6}$ is equivalent to a threefold rotation axis with a mirror plane perpendicular to the axis.

A centre of symmetry exists in a crystal if an imaginary line can be extended from any point on its surface through its centre and a similar point is present along the line equidistant from the centre. This is equivalent to $\bar{1}$, or inversion. There is a relatively simple procedure for recognizing a centre of symmetry in a well-formed crystal. With the crystal laid down on any face on a tabletop, the presence of a face of equal size and shape, but inverted, in a horizontal position at the top of the crystal proves the existence of a centre of symmetry. An imaginary mirror plane (or symmetry plane) can also be used to separate a crystal into halves. In a perfectly developed crystal, the halves are mirror images of one another.

Morphologically, crystals can be grouped into 32 crystal classes that represent the 32 possible symmetry elements and their combinations. These crystal classes, in turn, are grouped into six crystal systems. In decreasing order of overall symmetry content, beginning with the system with the highest and most complex crystal symmetry, they are isometric (or cubic), hexagonal, tetragonal, orthorhombic, monoclinic, and triclinic. (Many sources list seven crystal systems by dividing the hexagonal crystal system into two parts—trigonal and hexagonal).

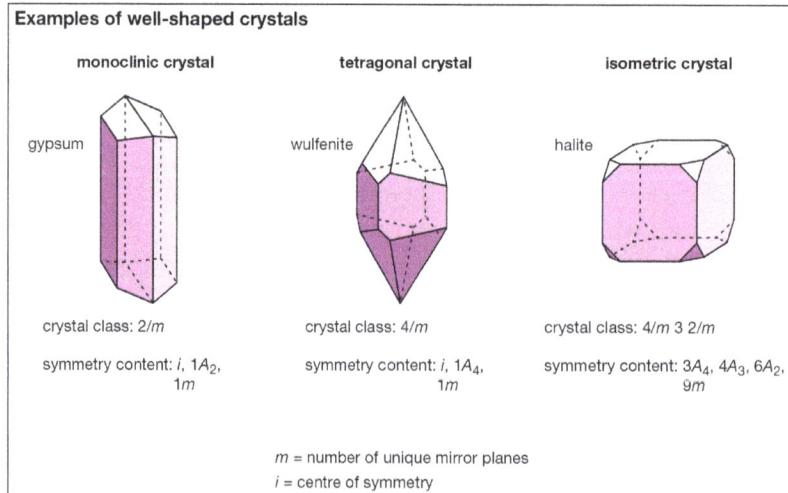

Examples of well-shaped crystals

monoclinic crystal	tetragonal crystal	isometric crystal
gypsum	wulfenite	halite
crystal class: 2/m	crystal class: 4/m	crystal class: 4/m 3 2/m
symmetry content: i, $1A_2$, $1m$	symmetry content: i, $1A_4$, $1m$	symmetry content: $3A_4$, $4A_3$, $6A_2$, $9m$

m = number of unique mirror planes
i = centre of symmetry

Examples of well-shaped crystals: Each of these examples of well-shaped crystals belongs to a different crystal class because of its overall symmetry content.

The systems may be described in terms of crystallographic axes used for reference. The c axis is normally the vertical axis. The isometric system exhibits three mutually perpendicular axes of equal length (a_1, a_2, and a_3). The orthorhombic and tetragonal systems also contain three mutually perpendicular axes; in the former system all the axes are of different lengths (a, b, and c), and in the latter system two axes are of equal length (a_1 and a_2) while the third (vertical) axis is either longer or shorter (c). The hexagonal system contains four axes: three equal-length axes (a_1, a_2, and a_3) intersect one another at 120° and lie in a plane that is perpendicular to the fourth (vertical) axis of a different length. Three axes of different lengths (a, b, and c) are present in both the monoclinic and triclinic systems. In the monoclinic system, two axes intersect one another at an oblique angle and lie in a plane perpendicular to the third axis; in the triclinic system, all axes intersect at oblique angles.

Twinning

If two or more crystals form a symmetrical intergrowth, they are referred to as twinned crystals. A new symmetry operation (called a twin element), which is lacking in a single untwinned crystal, relates the individual crystals in a twinned position. There are three twin elements that may relate the crystals of a twin: (1) reflection by a mirror plane (twin plane), (2) rotation about a crystal direction common to both (twin axis) with the angular rotation typically 180°, and (3) inversion about a point (twin centre). An instance of twinning is defined by a twin law that specifies the presence of a plane, an axis, or a centre of twinning. If a twin has three or more parts, it is referred to as a multiple, or repeated, twin.

Internal Structure

Examining Crystal Structures

The external morphology of a mineral is an expression of the fundamental internal architecture of a crystalline substance—i.e., its crystal structure. The crystal structure is the three-dimensional, regular (or ordered) arrangement of chemical units (atoms, ions, and anionic groups in inorganic materials; molecules in organic substances); these chemical units (referred to here as motifs) are repeated by various translational and symmetry operations. The morphology of crystals can be studied with the unaided eye in large well-developed crystals and has been historically examined in considerable detail by optical measurements of smaller well-formed crystals through the use of optical goniometers (instruments that measure the angles between crystal faces). The internal structure of crystalline materials, however, is revealed by a combination of X-ray, neutron, and electron diffraction techniques, supplemented by a variety of spectroscopicmethods, including infrared, optical, Mössbauer, and resonance-techniques. These methods, used singly or in combination, provide a quantitative three-dimensional reconstruction of the location of the atoms (or ions), the chemical bond types and their positions, and the overall internal symmetry of the structure. The repeat distances in most inorganic structures and many of the atomic and ionic motif sizes are on the order of 1 to 10 angstroms (Å; 1 Å is equivalent to 10^{-8} cm or 3.94×10^{-9} inch) or 10 to 100 nanometres (nm; 1 nm is equivalent to 10^{-7} cm or 10 Å).

Space Groups

Symmetry elements that are observable in the external morphology of crystals, such as rotation and rotoinversion axes, mirror planes, and a centre of symmetry, also are present in their internal atomic structure. In addition to these symmetry elements, there are translations and symmetry operations combined with translations. (Translation is the operation in which a motif is repeated in a linear pattern at intervals that are equal to the translation distance [commonly on the 1 to 10 Å level]). Two examples of translational symmetry elements are screw axes (combining rotation and translation) and glide planes (combining mirroring and translation). The internal translation distances are exceedingly small and can be seen directly only by very high-magnification electron beam techniques, as used in a transmission electron microscope, at magnifications of about 600,000 times.

When all possible combinations of translational elements compatible with the 32 crystal classes (also known as point groups) are considered, one arrives at 230 possible ways in which translations, translational symmetry elements (screw axes and glide planes), and translation-free symmetry elements (rotation and rotoinversion axes and mirror planes) can be combined. These translation and symmetry groupings are known as the 230 space groups, representing the various ways in which motifs can be arranged in an ordered three-dimensional array. The symbolic representation of space groups is closely related to that of the Hermann-Mauguin notation of point groups.

Illustrating Crystal Structures

The external morphology of three-dimensional arrangement of crystal structures may be presented on a two-dimensional page or within a computer simulation. Another common method of illustration involves projecting the crystal structure onto a planar surface. The high-temperature form of silicon dioxide (SiO_2) known as tridymite may be represented this way; however, the structural motif units in this case are SiO_4 tetrahedrons composed of a silicon atom surrounded by four oxygenatoms. To further aid the visualization of complex crystal structures within the physical world, three-dimensional physical models of such structures can be built or obtained commercially. Models of this sort reproduce the internal atomic arrangement on an enormously enlarged scale (e.g., one angstrom might be represented by one centimetre).

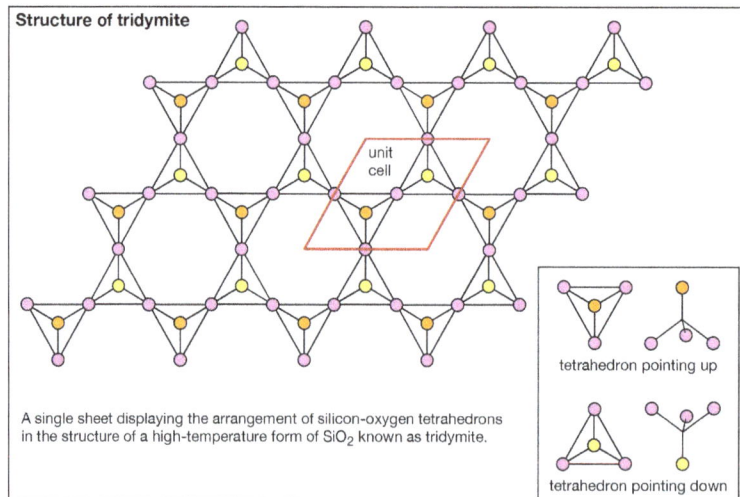

Silicon-oxygen tetrahedrons: Single sheet displaying the arrangement of the silicon-oxygen tetrahedrons in the structure of a high-temperature form of SiO_2 known as tridymite.

Polymorphism

Polymorphism is the ability of a specific chemical composition to crystallize in more than one form. This generally occurs as a response to changes in temperature or pressure or both. The different structures of such a chemical substance are called polymorphic forms, or polymorphs. For example, the element carbon (C) occurs in nature in two different polymorphic forms, depending on the external (pressure and temperature) conditions. These forms are graphite, with a hexagonal structure, and diamond, with an isometric structure. The composition FeS_2 occurs most commonly as pyrite, with an isometric structure, but it is also found as marcasite, which has an orthorhombic internal arrangement. The composition SiO_2 is found in a large number of polymorphs, among them quartz, tridymite, cristobalite, coesite, and stishovite. The stability field (conditions under which a mineral is stable) of these SiO_2 polymorphs can be expressed in a stability diagram, with the external parameters of temperature and pressure as the two axes. In the general quartz field, there is additional polymorphism leading to the notation of high quartz and low quartz, each form having a slightly different internal structure. Cristobaliteand tridymite are the high-temperature forms of SiO_2, and indeed these SiO_2 polymorphs occur in high-temperature lava flows. The high-pressure forms of SiO_2 are coesite and stishovite, and these can be found in meteorite craters, formed as a result of high explosive pressures upon quartz-rich sandstones, and in very deep-seated rock formations, as from Earth's upper mantle or very deep in subduction zones.

Pyrite: Pyrite from Navajun, Spain.

Chemical Composition

The chemical composition of a mineral is of fundamental importance because its properties greatly depend on it. Such properties, however, are determined not only by the chemical composition but also by the geometry of the constituent atoms and ions and by the nature of the electrical forces that bind them. Thus, for a complete understanding of minerals, their internal structure, chemistry, and bond types must be considered.

Various analytical techniques may be employed to obtain the chemical composition of a mineral. Quantitative chemical analyses mainly use so-called wet analytical methods (e.g., dissolution in acid, flame tests, and other classic techniques of bench chemistry that rely on observation), in which the mineral sample is first dissolved. Various compounds are then precipitated from the solution, which are weighed to obtain a gravimetric analysis. A number of analytical procedures have been introduced that provide faster but somewhat less accurate results. Most analyses use instrumental methods such as optical emission, X-ray fluorescence, atomic absorption spectroscopy, and electron microprobe analysis. Relatively well-established error ranges have been documented for these methods, and samples must be prepared in a specific manner for each technique. A distinct advantage of wet analytical procedures is that they make it possible to determine quantitatively the oxidation states of positively charged atoms, called cations (e.g., Fe^{2+} versus Fe^{3+}), and to ascertain the amount of water in hydrous minerals. It is more difficult to provide this type of information with instrumental techniques.

To ensure an accurate chemical analysis, the selected sample, which might include several minerals, is often made into a thin section (a section of rock less than 1 mm thick cemented for study between clear glass plates). To reduce the effect of the impurities, an instrumental technique, such as electron microprobe analysis, is commonly employed. In this method, quantitative analysis in situ may be performed on mineral grains only 1 micrometre (10^{-4} centimetre) in diameter.

Mineral Formulas

Elements may exist in the native (uncombined) state, in which case their formulas are simply their chemical symbols: gold (Au), carbon (C) in its polymorphic form of diamond, and sulfur (S) are common examples. Most minerals, however, occur as compounds consisting of two or more elements; their formulas are obtained from quantitative chemical analyses and indicate the relative proportions of the constituent elements. The formula of sphalerite, ZnS, reflects a one-to-one ratio between atoms of zincand those of sulfur. In bornite (Cu_5FeS_4), there are five atoms of copper (Cu), one atom of iron (Fe), and four atoms of sulfur. There exist relatively few minerals with constant composition; notable examples include quartz (SiO_2) and kyanite (Al_2SiO_5). Minerals of this sort are termed pure substances. Most minerals display considerable variation in the ions that occupy specific atomic sites within their structure. For example, the iron content of rhodochrosite ($MnCO_3$) may vary over a wide range. As ferrous iron (Fe^{2+}) substitutes for manganese cations (Mn^{2+}) in the rhodochrosite structure, the formula for the mineral might be given in more general terms—namely, $(Mn, Fe)CO_3$. The amounts of manganese and iron are variable, but the ratio of the cation to the negatively charged anionic group remains fixed at one Mn^{2+} or Fe^{2+} atom to one CO_3 group.

Sphalerite: Sphalerite, a mineral that is the principal ore of zinc.

Compositional Variation

Most minerals exhibit a considerable range in chemical composition. Such variation results from the replacement of one ion or ionic group by another in a particular structure. This phenomenon is termed ionic substitution, or solid solution. Three types of solid solution are possible, and these may be described in terms of their corresponding mechanisms—namely, substitutional, interstitial, and omission.

Substitutional solid solution is the most common variety. For example, in the carbonate mineral rhodochrosite ($MnCO_3$), Fe^{2+} may substitute for Mn^{2+} in its atomic site in the structure.

The degree of substitution may be influenced by various factors, with the size of the ion being the most important. Ions of two different elements can freely replace one another only if their ionic radii differ by approximately 15 percent or less. Limited substitution can occur if the radii differ by 15 to 30 percent, and a difference of more than 30 percent makes substitution unlikely. These limits, calculated from empirical data, are only approximate.

The temperature at which crystals grow also plays a significant role in determining the extent of ionic substitution. The higher the temperature, the more extensive is the thermal disorder in the crystal structure and the less exacting are the spatial requirements. As a result, ionic substitution that could not have occurred in crystals grown at low temperatures may be present in those grown at higher ones. The high-temperature form of $KAlSi_3O_8$ (sanidine), for example, can accommodate more sodium (Na) in place of potassium (K) than can microcline, its low-temperature counterpart.

An additional factor affecting ionic substitution is the maintenance of a balance between the positive and negative charges in the structure. Replacement of a monovalent ion (e.g., Na^+, a sodium cation) by a divalent ion (e.g., Ca^{2+}, a calcium cation) requires further substitutions to keep the structure electrically neutral.

Simple cationic or anionic substitutions are the most basic types of substitutional solid solution. A simple cationic substitution can be represented in a compound of the general form A^+X^- in which cation B^+ replaces in part or in total cation A^+. Both cations in this example have the same valence (+1), as in the substitution of K^+ (potassium ions) for Na^+ (sodium ions) in the NaCl (sodium chloride) structure. Similarly, the substitution of anion X^- by Y^- in an A^+X^- compound represents a simple anionic substitution; this is exemplified by the replacement of Cl^- (chlorineions) with Br^- (bromine ions) in the structure of KCl (potassium chloride). A complete solid-solution series involves the substitution in one or more atomic sites of one element for another that ranges over all possible compositions and is defined in terms of two end-members. For example, the two end-members of olivine [$(Mg, Fe)_2SiO_4$], forsterite (Mg_2SiO_4) and fayalite (Fe_2SiO_4), define a complete solid-solution series (called the forsterite-fayalite series) in which magnesium cations (Mg^{2+}) are replaced partially or totally by Fe^{2+}.

In some instances, a cation B^{3+} may replace some A^{2+} of compound $A^{2+}X^{2-}$. So that the compound will remain neutral, an equal amount of A^{2+} must concurrently be replaced by a third cation, C^+. This is given in equation form as $2A^{2+} \leftrightarrow B^{3+} + C^+$; the positive charge on each side is the same. Substitutions such as this are termed coupled substitutions. The plagioclase feldspar series exhibits complete solid solution, in the form of coupled substitutions, between its two end-members, albite ($NaAlSi_3O_8$) and anorthite ($CaAl_2Si_2O_8$). Every atomic substitution of Na^+ by Ca^{2+} is accompanied by the replacement of a silicon cation (Si^{4+}) by an aluminumcation (Al^{3+}), thereby maintaining electrical neutrality: $Na^+ + Si^{4+} \leftrightarrow Ca^{2+} + Al^{3+}$.

The second major type of ionic substitution is interstitial solid solution, or interstitial substitution. It takes place when atoms, ions, or molecules fill the interstices (voids) found between the atoms, ions, or ionic groups of a crystal structure. The interstices may take the form of channel-like cavities in certain crystals, such as the ring silicate beryl ($Be_3Al_2Si_6O_{18}$). Potassium, rubidium (Rb), cesium (Cs), and water, as well as helium (He), are some of the large ions and gases found in the tubular voids of beryl.

The least common type of solid solution is omission solid solution, in which a crystal contains one or more atomic sites that are not completely filled. The best-known example is exhibited by pyrrhotite ($Fe_{1-x}S$). In this mineral, each iron atom is surrounded by six neighbouring sulfur atoms. If every iron site in pyrrhotite were occupied by ferrous iron, its formula would be FeS. There are, however, varying percentages of vacancy in the iron site, so that the formula is given as Fe_6S_7 through $Fe_{11}S_{12}$, the latter being very near to pure FeS. The formula for pyrrhotite is normally written as $Fe_{1-x}S$, with x ranging from 0 to 0.2. It is one of the minerals referred to as a defect structure, because it has a structural site that is not completely occupied.

Chemical Bonding

Electrical forces are responsible for the chemical bonding of atoms, ions, and ionic groups that constitute crystalline solids. The physical and chemical properties of minerals are attributable for the most part to the types and strengths of these binding forces; hardness, cleavage, fusibility, electrical and thermal conductivity, and the coefficient of thermal expansion are examples of such properties. On the whole, the hardness and melting point of a crystal increase proportionally with the strength of the bond, while its coefficient of thermal expansion decreases. The extremely strong forces that link the carbon atoms of diamond, for instance, are responsible for its distinct hardness. Periclase (MgO) and halite (NaCl) have similar structures; however, periclase has a melting point of 2,800 °C (5,072 °F) whereas halite melts at 801 °C (1,474 °F). This discrepancy reflects the difference in the bond strength of the two minerals: since the atoms of periclase are joined by a stronger electrical force, a greater amount of heat is needed to separate them.

The electrical forces, called chemical bonds, can be divided into five types: ionic, covalent, metallic, van der Waals, and hydrogen bonds. Classification in this manner is largely one of expediency; the chemical bonds in a given mineral may in fact possess characteristics of more than one bond type. For example, the forces that link the silicon and oxygen atoms in quartz exhibit in nearly equal amount the characteristics of both ionic and covalent bonds. As stated above, the electrical interaction between the atoms of a crystal determines its physical and chemical properties. Thus, classifying minerals according to their electrical forces will cause those species with similar properties to be grouped together. This fact justifies classification by bond type.

Ionic Bonds

Atoms have a tendency to gain or lose electrons so that their outer orbitals become stable; this is normally accomplished by these orbitals being filled with the maximum allowed number of valence electrons. Metallic sodium, for example, has one valence electron in its outer orbital; it becomes ionized by readily losing this electron and exists as the cation Na^+. Conversely, chlorine gains an electron to complete its outer orbital, thereby forming the anion Cl^-. In the mineral halite, NaCl (common, or rock, salt), the chemical bonding that holds the Na^+ and Cl^- ions together is the attraction between the two opposite charges. This bonding mechanism is referred to as ionic, or electrovalent.

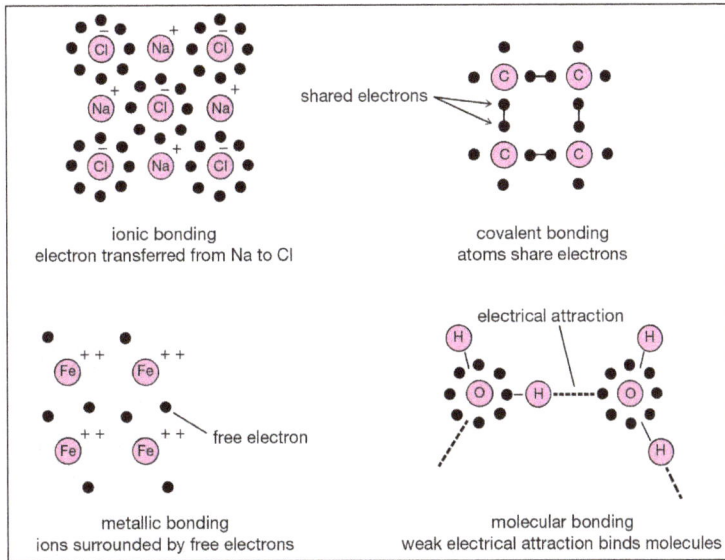

Crystal bonding: Different types of bonding in crystals.

Ionically bonded crystals typically display moderate hardness and specific gravity, rather high melting points, and poor thermal and electrical conductivity. The electrostatic charge of an ion is evenly distributed over its surface, and so a cation tends to become surrounded with the maximum number of anions that can be arranged around it. Since ionic bonding is non-directional, crystals bonded in this manner normally displays high symmetry.

Covalent Bonds

Chlorine readily gains an electron to achieve a stable electron configuration. An incomplete outer orbital places a chlorine atom in a highly reactive state, so it attempts to combine with nearly any atom in its proximity. Because its closest neighbour is usually another chlorine atom, the two may bond together by sharing one pair of electrons. As a result of this extremely strong bond, each chlorine atom enters a stable state.

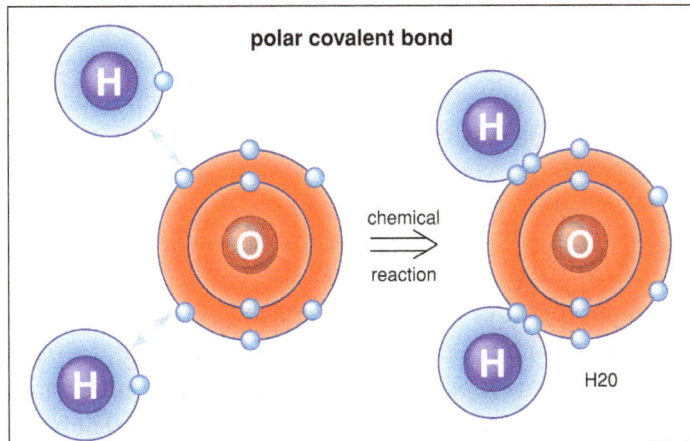

Polar covalent bond

In figure, in polar covalent bonds, such as that between hydrogen and oxygen atoms, the electrons are not transferred from one atom to the other as they are in an ionic bond. Instead, some outer electrons merely spend more time in the vicinity of the other atom. The effect of this orbital distortion is to induce regional net charges that hold the atoms together, such as in water molecules.

The electron-sharing, or covalent, bond is the strongest of all chemical bond types. Minerals bonded in this manner display general insolubility, great stability, and a high melting point. Crystals of covalently bonded minerals tend to exhibit lower symmetry than their ionic counterparts because the covalent bond is highly directional, localized in the vicinity of the shared electrons.

The Cl_2 molecules formed by linking two neighbouring chlorine atoms are stable and do not combine with other molecules. Atoms of some elements, however, have more than one electron in the outer orbital and thus may bond to several neighbouring atoms to form groups, which in turn may join together in larger combinations. Carbon, in the polymorphic form of diamond, is a good example of this type of covalent bonding. There are four valence electrons in a carbon atom, so that each atom bonds with four others in a stable tetrahedral configuration. A continuous network is formed by the linkage of every carbon atom in this manner. The rigid diamond structure results from the strong localization of the bond energy in the vicinity of the shared electrons; this makes diamond the hardest of all natural substances. Diamond does not conduct electricity, because all the valence electrons of its constituent atoms are shared to form bonds and therefore are not mobile.

Metallic Bonds

Bonding in metals is distinct from that in their salts, as reflected in the significant differences between the properties of the two groups. In contrast to salts, metals display high plasticity, tenacity, ductility, and conductivity. Many are characterized by lower hardness and have higher melting and boiling points than, for example, covalently bonded materials. All these properties result from a metallic bonding mechanism that can be envisioned as a collection of positively charged ions immersed in a cloud of valence electrons. The attraction between the cations and the electrons holds a crystal together. The electrons are not bound to any particular cation and are thus free to move throughout the structure. In fact, in the metals sodium, cesium, rubidium, and potassium, the radiant energy of light can cause electrons to be removed from their surfaces entirely. (This result is known as the photoelectric effect). Electron mobility is responsible for the ability of metals to conduct heat and electricity. The native metals are the only minerals to exhibit pure metallic bonding.

Van Der Waals Bonds

Neutral molecules may be held together by a weak electric force known as the van der Waals bond. It results from the distortion of a molecule so that a small positive charge develops on one end and a corresponding negative charge develops on the other. A similar effect is induced in neighbouring molecules, and this dipole effect propagates throughout the entire structure. An attractive force is then formed between oppositely charged ends of the dipoles. Van der Waals bonding is common in gases and organic liquids and solids, but it is rare in minerals. Its presence in a mineral defines a weak area with good cleavage and low hardness. In graphite, carbon atoms lie in covalently bonded sheets with van der Waals forces acting between the layers.

Hydrogen Bonds

In addition to the four major bond types described above, there is an interaction called hydrogen bonding. This takes place when a hydrogen atom, bonded to an electronegative atom such as oxygen, fluorine, or nitrogen, is also attracted to the negative end of a neighbouring molecule. A strong dipole-dipole interaction is produced, forming a bond between the two molecules. Hydrogen bonding is common in hydroxides and in many of the layer silicates—e.g., micas and clay minerals.

Physical Properties

The physical properties of minerals are the direct result of the structural and chemical characteristics of the minerals. Some properties can be determined by inspection of a hand specimen or by relatively simple tests on such a specimen. Others, such as those determined by optical and X-ray diffraction techniques require special and often sophisticated equipment and may involve elaborate sample preparation.

Crystal Habit and Crystal Aggregation

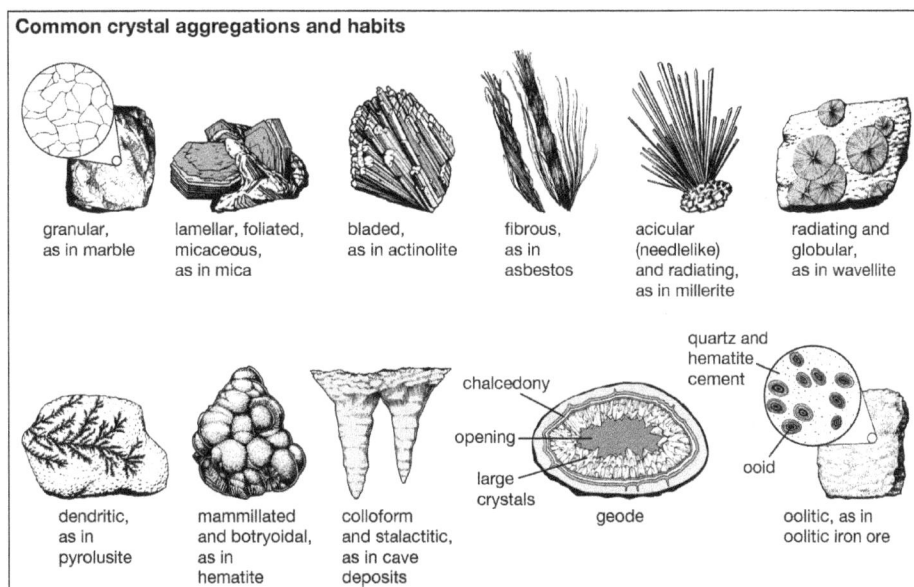

Common crystals: Common crystal aggregations and habits.

The external shape (habit) of well-developed crystals can be visually studied and classified according to the various crystal systems that span the 32 crystal classes. The majorities of crystal occurrences, however, is not part of well-formed single crystals but are found as crystals grown together in aggregates. Examples of some descriptive terms for such aggregations are given here: granular, an intergrowth of mineral grains of approximately the same size; lamellar, flat, platelike individuals arranged in layers; bladed, elongated crystals flattened like a knife blade; fibrous, an aggregate of slender fibres, parallel or radiating; acicular, slender, needlelike crystals; radiating, individuals forming starlike or circular groups; globular, radiating individuals forming small spherical or hemispherical groups; dendritic, in slender divergent branches, somewhat plantlike; mammillary, large smoothly rounded, masses resembling mammae, formed by radiating crystals; botryoidal, globular forms resembling a bunch of grapes; colloform, spherical forms composed of radiating individuals without regard to size (this includes botryoidal, reniform, and mammillary forms); stalactitic, pendant cylinders or cones resembling icicles; concentric, roughly spherical layers arranged about a common centre, as in agate and in geodes; geode, a partially filled rock cavity lined by mineral material (geodes may be banded as in agate owing to successive depositions of material, and the inner surface is often covered with projecting crystals); and oolitic, an assemblage consisting of small spheres.

Cleavage and Fracture

Both these properties represent the reaction of a mineral to an external force. Cleavage is breakage along planar surfaces, which are parallel to possible external faces on the crystal. It results from the tendency of some minerals to split in certain directions that are structurally weaker than others. Some crystals exhibit well-developed cleavage, as seen by the planar cleavage in mica; perfect cleavage of this sort is characterized by smooth, shiny surfaces. In other minerals, such as quartz, cleavage is absent. Quality and direction are the general characteristics used to describe cleavage. Quality is expressed as perfect, good, fair, and so forth; cleavage directions of a crystal are consistent with its overall symmetry.

Augite: Photomicrograph of various pyroxene minerals in thin sections (illuminated with polarized light). An augite phenocryst (large individual gray crystal) appears in basalt lava, showing characteristic basal octagonal form and square-segmentation cleavage.

Some crystals do not usually break in any particular direction, reflecting roughly equal bond strengths throughout the crystal structure. Breakage in such minerals is known as fracture. The term *conchoidal* is used to describe fracture with smooth, curved surfaces that resemble the interior of a seashell; it is commonly observed in quartz and glass. Splintery fracture is breakage into elongated fragments like splinters of wood, while hackly fracture is breakage along jagged surfaces.

Lustre

The term *lustre* refers to the general appearance of a mineral surface in reflected light. The main types of lustre, metallic and nonmetallic, are distinguished easily by the human eye after some practice, but the difference between them cannot be quantified and is rather difficult to describe. Metallic refers to the lustre of an untarnished metallic surface such as gold, silver, copper, or steel. These materials are opaque to light; none passes through even at thin edges. Pyrite (FeS_2), chalcopyrite ($CuFeS_2$), and galena (PbS) are common minerals that have metallic lustre. Nonmetallic lustre is generally exhibited by light-coloured minerals that transmit light, either through thick portions or at least through their edges. The following terms are used to distinguish the lustre of nonmetallic minerals: vitreous, having the lustre of a piece of broken glass (this is commonly seen in quartz and many other nonmetallic minerals); resinous, having the lustre of a piece of resin (this is common in sphalerite [ZnS]); pearly, having the lustre of mother-of-pearl (i.e., an iridescent pearl-like lustre characteristic of mineral surfaces that are parallel to well-developed cleavage planes; the cleavage surface of talc [$Mg_3Si_4O_{10}(OH)_2$] may show pearly lustre); greasy, having the appearance of being covered with a thin layer of oil (such lustre results from the scattering of light by a microscopically rough surface; some nepheline [$(Na, K)AlSiO_4$] and milky quartz may exhibit this); silky, descriptive of the lustre of a skein of silk or a piece of satin and characteristic of some minerals in fibrous aggregates (examples are fibrous gypsum [$CaSO_4 \cdot 2H_2O$], known as satin spar, and chrysotile asbestos [$Mg_3Si_2O_5(OH)_4$]); and adamantine, having the brilliant lustre of diamond, exhibited by minerals with a high refractive indexcomparable to diamond and which as such refract light as strongly as the latter (examples are cerussite [$PbCO_3$] and anglesite [$PbSO_4$]).

Galena: Galena is the most common mineral that contains lead.

Colour

Minerals occur in a great variety of colours. Because colour varies not only from one mineral to another but also within the same mineral (or mineral group), the observer must learn in which minerals it is a constant property and can thus be relied on as a distinguishing criterion. Most minerals that have a metallic lustre very little in colour, but nonmetallic minerals can demonstrate wide variance. Although the colour of a freshly broken surface of a metallic mineral is often highly diagnostic, this same mineral may become tarnished with time. Such tarnish may dull minerals such as galena (PbS), which has a bright bluish lead-gray colour on a fresh surface but may become dull upon long exposure to air. Bornite (Cu_5FeS_4), which on a freshly broken surface has a brownish bronze colour, may be so highly tarnished on an older surface that it shows variegated purples and blues; hence, it is called peacock ore. In other words, in the identification of minerals with a metallic lustre, it is important for the observer to have a freshly broken surface for accurate determination of colour.

Bornite

A few minerals with nonmetallic lustre display a constant colour that can be used as a truly diagnostic property. Examples are malachite, which is green; azurite, which is blue; rhodonite, which is pink; turquoise, which gives its name to the colour turquoise, a greenish blue to blue-green; and sulfur, which is yellow. Many nonmetallic minerals have a relatively narrow range of colours, although some have an unusually wide range. Members of the plagioclase feldspar series range from almost pure white in albitethrough light gray to darker gray toward the anorthite end-member. Most common garnets show various shades of red to red-brown to brown. Members of the monoclinic pyroxene group range from almost white in pure diopside to light green in diopside containing a small amount of ironas a substitute for magnesium in the structure through dark green in hedenbergite to almost black in many augites. Members of the orthopyroxene series (enstatite to orthoferrosilite) range from light beige to darker brown. On the other hand, tourmaline may show many colours (red, blue, green, brown, and black) as well as distinct colour zonation, from colourless through pink to green, within a single crystal. Similarly, numerous gem minerals such as corundum, beryl, and quartz occur in many colours; the gemstones cut from them are given varietal names. In short, in nonmetallic minerals of various kinds, colour is a helpful, though not a truly diagnostic (and therefore unique), property.

Hardness

Hardness (H) is the resistance of a mineral to scratching. It is a property by which minerals may be described relative to a standard scale of 10 minerals known as the Mohs scale of hardness. The degree of hardness is determined by observing the comparative ease or difficulty with which one mineral is scratched by another or by a steel tool. For measuring the hardness of a mineral, several common objects that can be used for scratching are helpful, such as a fingernail, a copper coin, a steel pocketknife, glass plate or window glass, the steel of a needle, and a streakplate (an unglazed black or white porcelain surface).

Because there is a general link between hardness and chemical composition, these generalizations can be made:

1. Most hydrous minerals are relatively soft (H < 5).

2. Halides, carbonates, sulfates, and phosphates also are relatively soft (H < 5.5).

3. Most sulfides are relatively soft (H < 5), with marcasite and pyrite being examples of exceptions (H < 6 to 6.5).

4. Most anhydrous oxides and silicates are hard (H > 5.5).

Because hardness is a highly diagnostic property in mineral identification, most determinative tables use relative hardness as a sorting parameter.

Tenacity

Several mineral properties that depend on the cohesive force between atoms (and ions) in mineral structures are grouped under tenacity. A mineral's tenacity can be described by the following terms: malleable, capable of being flattened under the blows of a hammer into thin sheets without breaking or crumbling into fragments (most of the native elements show various degrees of malleability, but particularly gold, silver, and copper); sectile, capable of being severed by the smooth cut of a knife (copper, silver, and gold are sectile); ductile, capable of being drawn into the form of a wire (gold, silver, and copper exhibit this property); flexible, bending easily and staying bent after the pressure is removed (talc is flexible); brittle, showing little or no resistance to breakage, and as such separating into fragments under the blow of a hammer or when cut by a knife (most silicate minerals are brittle); and elastic, capable of being bent or pulled out of shape but returning to the original form when relieved (mica is elastic).

Gold leaf: Burnishing gold leaf.

Specific Gravity

Specific gravity (G) is defined as the ratio between the weight of a substance and the weight of an equal volume of water at 4 °C (39 °F). Thus a mineral with a specific gravity of 2 weighs twice as much as the same volume of water. Since it is a ratio, specific gravity has no units.

The specific gravity of a mineral depends on the atomic weights of all its constituent elements and the manner in which the atoms (and ions) are packed together. In mineral series whose species have essentially identical structures, those composed of elements with higher atomic weight have higher specific gravities. If two minerals (as in the two polymorphs of carbon, namely graphite and diamond) have the same chemical composition, the difference in specific gravity reflects variation in internal packing of the atoms or ions (diamond, with a G of 3.51, has a more densely packed structure than graphite, with a G of 2.23).

Measurement of the specific gravity of a mineral specimen requires the use of a special apparatus. An estimate of the value, however, can be obtained by simply testing how heavy a specimen feels. Most people, from everyday experience, have developed a sense of relative weights for even such objects as nonmetallic and metallic minerals. For example, borax (G = 1.7) seems light for a nonmetallic mineral, whereas anglesite (G = 6.4) feels heavy. Average specific gravity reflects what a nonmetallic or metallic mineral of a given size should weigh. The average specific gravity for nonmetallic minerals falls between 2.65 and 2.75, which is seen in the range of values for quartz (G = 2.65), feldspar (G = 2.60 to 2.75), and calcite (G = 2.72). For metallic minerals, graphite (G = 2.23) feels light, while silver (G = 10.5) seems heavy. The average specific gravity for metallic minerals is approximately 5.0, the value for pyrite. With practice using specimens of known specific gravity, a person can develop the ability to distinguish between minerals that have comparatively small differences in specific gravity by merely lifting them.

Although an approximate assessment of specific gravity can be obtained by the hefting of a hand specimen of a specific monomineral, an accurate measurement can only be achieved by using a specific gravity balance. An example of such an instrument is the Jolly balance, which provides numerical values for a small mineral specimen (or fragment) in air as well as in water. Such accurate measurements are highly diagnostic and can greatly aid in the identification of an unknown mineral sample.

Magnetism

Only two minerals exhibit readily observed magnetism: magnetite (Fe_3O_4), which is strongly attracted to a hand magnet, and pyrrhotite ($Fe_{1-x}S$), which typically shows a weaker magnetic reaction. Ferromagnetic is a term that refers to materials that exhibit strong magnetic attraction when subjected to a magnetic field. Materials that show only a weak magnetic response in a strong applied magnetic field are known as paramagnetic. Those materials that are repelled by an applied magnetic force are known as diamagnetic. Because minerals display a wide range of slightly different magnetic properties, they can be separated from each other by an electromagnet. Such magnetic separation is a common procedure both in the laboratory and on a commercial scale.

Fluorescence

Some minerals, when exposed to ultraviolet light, will emit visible light during irradiation; this is known as fluorescence. Some minerals fluoresce only in shortwave ultraviolet light, others only in longwave ultraviolet light, and still others in either situation. Both the colour and intensity of the emitted light vary significantly with the wavelengths of ultraviolet light. Due to the unpredictable nature of fluorescence, some specimens of a mineral manifest it, while other seemingly similar specimens, even those from the same geographic area, do not. Some minerals that may exhibit fluorescence are fluorite, scheelite, calcite, scapolite, willemite, and autunite. Specimens of willemite and calcite from the Franklin district of New Jersey in the United States may show brilliant fluorescent colours.

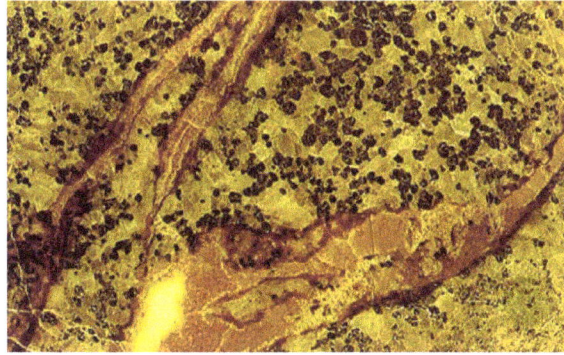

Franklinite and willemite: In ordinary light, the willemite grains and veins of this specimen are brownish gold and the franklinite grains are dark brown.

Solubility in Hydrochloric Acid

The positive identification of carbonate minerals is aided greatly by the fact that the carbon-oxygen bond of the CO_3 group in carbonates becomes unstable and breaks down in the presence of hydrogen ions (H^+) available in acids. This is expressed by the reaction $2H^+ + CO^{2-}_3 \rightarrow H_2O + CO_2$, which is the basis for the so-called fizz test with dilute hydrochloric acid (HCl). Calcite, aragonite, witherite, and strontianite, as well as copper carbonates, show bubbling, or effervescence, when a drop of dilute hydrochloric acid is placed on the mineral. This "fizz" is due to the release of carbon dioxide (CO_2). Other carbonates such as dolomite, rhodochrosite, magnesite, and siderite will show slow effervescence when acid is applied to powdered minerals or moderate effervescence only in hot hydrochloric acid.

Radioactivity

Minerals containing uranium (U) and thorium (Th) continually undergo decay reactions in which radioactive isotopes of uranium and thorium form various daughter elements and also release energy in the form of alpha and beta particles and gamma radiation. The radiation produced can be measured in the laboratory or in the field using a Geiger counter or a scintillation counter. A radiation counter therefore is helpful in identifying uranium- and thorium-containing minerals, such as uraninite, pitchblende, thorianite, and autunite. Several rock-forming minerals contain enough radioactive elements to permit the determination of the time elapsed since the radioactive material was incorporated into the mineral.

Classification of Minerals

The broadest divisions of the classification used in the present discussion are:

1. Native elements,

2. Sulfides,

3. Sulfosalts,

4. Oxides and hydroxides,

5. Halides,

6. Carbonates,

7. Nitrates,

8. Borates,

9. Sulfates,

10. Phosphates,

11. Silicates.

Native Elements

Apart from the free gases in Earth's atmosphere, some 20 elements occur in nature in a pure (i.e., uncombined) or nearly pure form. Known as the native elements, they are partitioned into three families: metals, semimetals, and nonmetals. The most common native metals, which are characterized by simple crystal structures, make up three groups: the goldgroup, consisting of gold, silver, copper, and lead; the platinum group, composed of platinum, palladium, iridium, and osmium; and the irongroup, containing iron and nickel-iron. Mercury, tantalum, tin, and zinc are other metals that have been found in the native state. The native semimetals are divided into two isostructural groups (those whose members share a common structure type): (1) antimony, arsenic, and bismuth, with the latter two being more common in nature, and (2) the rather uncommon selenium and tellurium. Carbon, in the form of diamondand graphite, and sulfur are the most important native nonmetals.

Native elements	
Metals	
Gold group	
Gold	Au
Silver	Ag
Copper	Cu
Platinum group	
Platinum	Pt
Iron group	
Iron	Fe
(Kamacite	Fe, ni)
(Taenite	Fe, ni)
Semimetals	
Arsenic group	
Arsenic	As
Bismuth	Bi

Native elements	
Nonmetals	
Sulfur	S
Diamond	C
Graphite	C

Metals

Gold, silver, and copper are members of the same group (column) in the periodic table of elements and therefore have similar chemical properties. In the uncombined state, their atoms are joined by the fairly weak metallic bond. These minerals share a common structure type, and their atoms are positioned in a simple cubic closest-packed arrangement. Gold and silver both have an atomic radius of 1.44 angstroms (Å), or 1.44×10^{-7} millimetre, which enables complete solid solution to take place between them. The radius of copper is significantly smaller (1.28 Å), and as such copper substitutes only to a limited extent in gold and silver. Likewise, native copper contains only trace amounts of gold and silver in its structure.

Ore: A sample of gold ore.

Because of their similar crystal structure, the members of the gold group display similar physical properties. All are rather soft, ductile (capable of being drawn into wire), malleable (capable of being shaped by a hammeror rollers), and sectile (capable of being cut smoothly by a knife or other instrument); gold, silver, and copper serve as excellent conductors of electricity and heat and exhibit metallic lustre and hackly fracture (a type of fracture characterized by sharp jagged surfaces). These properties are attributable to their metallic bonding. The gold-group minerals crystallize in the isometric system and have high densities as a consequence of cubic closest packing.

In addition to the elements listed above, the platinum group also includes rare mineral alloys such as iridosmine. The members of this group are harder than the metals of the gold group and also have higher melting points.

The iron-group metals are isometric and have a simple cubic packed structure. Its members include pure iron, which is rarely found on the surface of Earth, and two species of nickel-iron (kamacite and taenite), which have been identified as common constituents of meteorites. Native iron has been found in basalts of Disko Island, Greenland and nickel-iron in Josephine and Jackson counties, Oregon. The atomic radii of iron and nickel are both approximately 1.24 Å, and so nickel is a frequent substitute for iron. Earth's core is thought to be composed largely of such iron-nickel alloys.

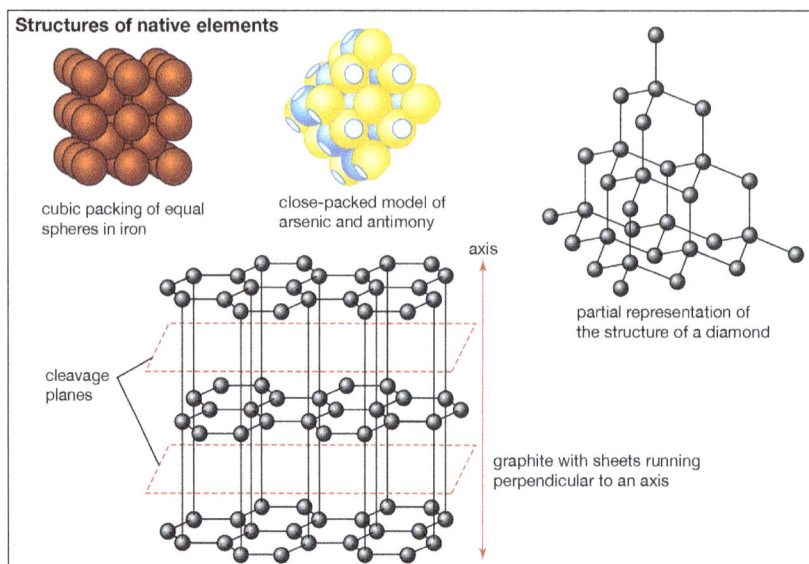

In figure, structures of some native elements. (A) Close-packed model of simple cubic packing of equal spheres, as shown by iron. Each sphere is surrounded by eight closest neighbours. (B) Close-packed model of the structure of arsenic and antimony. Flat areas represent overlap between adjoining atoms. (C) Partial representation of the structure of diamond. (D) The structure of graphite with sheets perpendicular to the c axis.

Semimetals

The semimetals antimony, arsenic, and bismuth have a structure type distinct from the simple-packed spheres of the metals. In these semimetals, each atom is positioned closer to three of its neighbouring atoms than to the rest. The structure of antimony and arsenic is composed of spheres that intersect along flat circular areas.

Arsenic (gray) with realgar (red) and orpiment (yellow).

The covalent character of the bonds joining the four closest atoms is linked to the electronegative nature of the semimetals, reflected by their position in the periodic table. Members of this group are fairly brittle, and they do not conduct heat and electricity nearly as well as the native metals. The bond type suggested by these properties is intermediate between metallic and covalent; it is consequently stronger and more directional than pure metallic bonding, resulting in crystals of lower symmetry.

Nonmetals

The native nonmetals diamond, fullerene, graphite, and sulfur are structurally distinct from the metals and semimetals. The structure of sulfur (atomic radius = 1.04 Å), usually orthorhombic in form, may contain limited solid solution by selenium (atomic radius = 1.16 Å).

The Hope diamond; in the Smithsonian Institution, Washington, D.C.

The polymorphs of carbon—graphite, fullerene, and diamond—display dissimilar structures, resulting in their differences in hardness and specific gravity. In diamond, each carbon atom is bonded covalently in a tetrahedral arrangement, producing a strongly bonded and exceedingly close-knit but not closest-packed structure. The carbon atoms of graphite, however, are arranged in six-membered rings in which each atom is surrounded by three close-by neighbours located at the vertices of an equilateral triangle. The rings are linked to form sheets, called graphene, that are separated by a distance exceeding one atomic diameter. Van der Waals forces act perpendicular to the sheets, offering a weak bond, which, in combination with the wide spacing, leads to perfect basal cleavage and easy gliding along the sheets. Fullerenes are found in meta-anthracite, in fulgurites, and in clays from the Cretaceous-Tertiary boundary in New Zealand, Spain, and Turkmenistan as well as in organic-rich layers near the Sudbury nickel mine of Canada.

Sulfides

This important class includes most of the ore minerals. The similar but rarer sulfarsenides are grouped here as well. Sulfide minerals consist of one or more metals combined with sulfur; sulfarsenides contain arsenic replacing some of the sulfur.

Sulfides are generally opaque and exhibit distinguishing colours and streaks. (Streak is the colour of a mineral's powder). The nonopaque varieties (e.g., cinnabar, realgar, and orpiment) possess high refractive indices, transmitting light only on the thin edges of a specimen.

Few broad generalizations can be made about the structures of sulfides, although these minerals can be classified into smaller groups according to similarities in structure. Ionic and covalent bonding are found in many sulfides, while metallic bonding is apparent in others as evidenced by their metal properties. The simplest and most symmetric sulfide structure is based on the architecture of the sodium chloride structure. A common sulfide mineral that crystallizes in this manner is the ore mineral of lead, galena. Its highly symmetric form consists of cubes modified by octahedral faces at their corners. The structure of the common sulfide pyrite (FeS_2) also is modeled after the sodium chloride type; a disulfide grouping is located in a position of coordination with six surrounding ferrous iron atoms. The high symmetry of this structure is reflected in the external morphology of pyrite. In another sulfide structure, sphalerite (ZnS), each zinc atom is surrounded by four sulfur atoms in a tetrahedral coordinating arrangement. In a derivative of this structure type, the chalcopyrite ($CuFeS_2$) structure, copper and iron ions can be thought of as having been regularly substituted in the zinc positions of the original sphalerite atomic arrangement.

Arsenopyrite ($FeAsS$) is a common sulfarsenide that occurs in many ore deposits. It is the chief source of the element arsenic.

Sulfosalts

There are approximately 100 species constituting the rather large and very diverse sulfosalt class of minerals. The sulfosalts differ notably from the sulfides and sulfarsenides with regard to the role of semimetals, such as arsenic (As) and antimony (Sb), in their structures. In the sulfarsenides, the semimetals substitute for some of the sulfur in the structure, while in the sulfosalts they are found instead in the metal site. For example, in the sulfarsenide arsenopyrite ($FeAsS$), the arsenic replaces sulfur in a marcasite- (FeS_2-) type structure. In contrast, the sulfosalt enargite (Cu_3AsS_4) contains arsenic in the metal position, coordinated to four sulfur atoms. A sulfosalt such as Cu_3AsS_4 may also be thought of as a double sulfide, $3Cu_2S \cdot As_2S_5$.

Oxides and Hydroxides

These classes consist of oxygen-bearing minerals; the oxides combine oxygen with one or more metals, while the hydroxides are characterized by hydroxyl $(OH)^-$ groups.

The oxides are further divided into two main types: simple and multiple. Simple oxides contain a single metal combined with oxygen in one of several possible metal : oxygen ratios (X:O): XO, X_2O, X_2O_3, etc. Ice, H_2O, is a simple oxide of the X_2O type that incorporates hydrogen as the cation. Although SiO_2 (quartz and its polymorphs) is the most commonly occurring oxide, it is discussed below in the section on silicates because its structure more closely resembles that of other silicon-oxygen compounds. Two nonequivalent metal sites (X and Y) characterize multiple oxides, which have the form XY_2O_4.

Unlike the minerals of the sulfide class, which exhibit ionic, covalent, and metallic bonding, oxide minerals generally display strong ionic bonding. They are relatively hard, dense, and refractory.

Oxides generally occur in small amounts in igneous and metamorphic rocks and also as pre-existing grains in sedimentary rocks. Several oxides have great economic value, including the principal ores of iron (hematiteand magnetite), chromium (chromite), manganese (pyrolusite, as well as the hydroxides, manganite and romanechite), tin (cassiterite), and uranium(uraninite).

Members of the hematite group are of the X_2O_3 type and have structures based on hexagonal closest packing of the oxygen atoms with octahedrally coordinated (surrounded by and bonded to six atoms) cations between them. Corundum and hematite share a common hexagonal architecture. In the ilmenite structure, iron and titanium occupy alternate Fe-O and Ti-O layers.

The XO_2- type oxides are divided into two groups. The first structure type, exemplified by rutile, contains cations in octahedral coordination with oxygen. The second resembles fluorite (CaF_2); each oxygen is bonded to four cations located at the corners of a fairly regular tetrahedron, and each cation lies within a cube at whose corners are eight oxygen atoms. This latter structure is exhibited by uranium, thorium, and cerium oxides, whose considerable importance arises from their roles in nuclear chemistry.

The spinel-group minerals have type XY_2O_4 and contain oxygen atoms in approximate cubic closest packing. The cations located within the oxygen framework are octahedrally (sixfold) and tetrahedrally (fourfold) coordinated with oxygen.

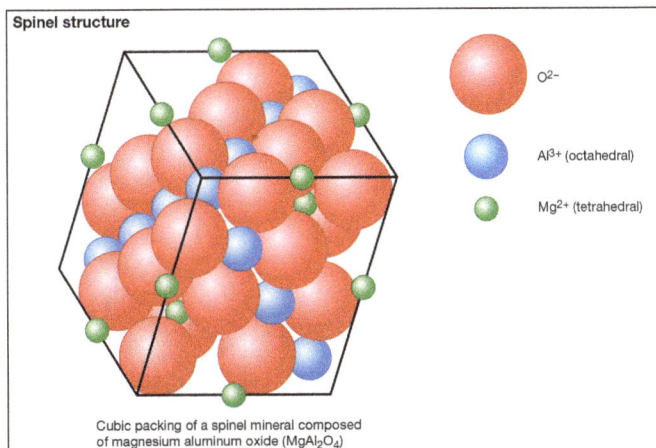

Spinel structure

O^{2-}

Al^{3+} (octahedral)

Mg^{2+} (tetrahedral)

Cubic packing of a spinel mineral composed of magnesium aluminum oxide ($MgAl_2O_4$)

An oxygen layer in the spinel ($MgAl_2O_4$) structure. The large circles represent oxygen in approximate cubic closest packing; the cation layers on each side of the oxygen layer are also shown.

The $(OH)^-$ group of the hydroxides generally results in structures with lower bond strengths than in the oxide minerals. The hydroxide minerals tend to be less dense than the oxides and also are not as hard. All hydroxides form at low temperatures and are found predominantly as weathering products, as, for example, from alteration in hydrothermal veins. Some common hydroxides are brucite $[Mg(OH)_2]$, manganite $[MnO \cdot OH]$, diaspore $[\alpha\text{-}AlO \cdot OH]$, and goethite $[\alpha\text{-}FeO \cdot OH]$. The ore of aluminum, bauxite, consists of a mixture of diaspore, boehmite ($\gamma\text{-}AlO \cdot OH$—a polymorph of diaspore), and gibbsite $[Al(OH)_3]$, plus iron oxides. Goethite is a common alteration product of iron-rich occurrences and is an iron ore in some localities.

Halides

Members of this class are distinguished by the large-sized anions of the halogens chlorine, bromine, iodine, and fluorine. The ions carry an electric charge of negative one and easily become distorted in the presence of strongly charged bodies. When associated with rather large, weakly polarizing cations of low charge, such as those of the alkali metals, both anions and cations take the form of nearly perfect spheres. Structures composed of these spheres exhibit the highest possible symmetry.

Pure ionic bonding is exemplified best in the isometric halides, for each spherical ion distributes its weak electrostatic charge over its entire surface. These halides manifest relatively low hardness and moderate-to-high melting points. In the solid state they are poor thermal and electric conductors, but when molten they conduct electricity well.

Halogen ions may also combine with smaller, more strongly polarizing cations than the alkali metal ions. Lower symmetry and a higher degree of covalent bonding prevail in these structures. Water and hydroxyl ions may enter the structure, as in atacamite [$Cu_2Cl(OH)_3$].

The halides consist of about 80 chemically related minerals with diversestructures and widely varied origins. The most common are halite (NaCl), sylvite (KCl), chlorargyrite (AgCl), cryolite (Na_3AlF_6), fluorite (CaF_2), and atacamite. No molecules are present among the arrangement of the ions in halite, a naturally occurring form of sodium chloride. Each cation and anion is in octahedral coordination with its six closest neighbours. The NaCl structure is found in the crystals of many XZ-type halides, including sylvite (KCl) and chlorargyrite (AgCl). Some sulfides and oxides of XZ type crystallize in this structure type as well—for example, galena (PbS), alabandite (MnS), and periclase (MgO).

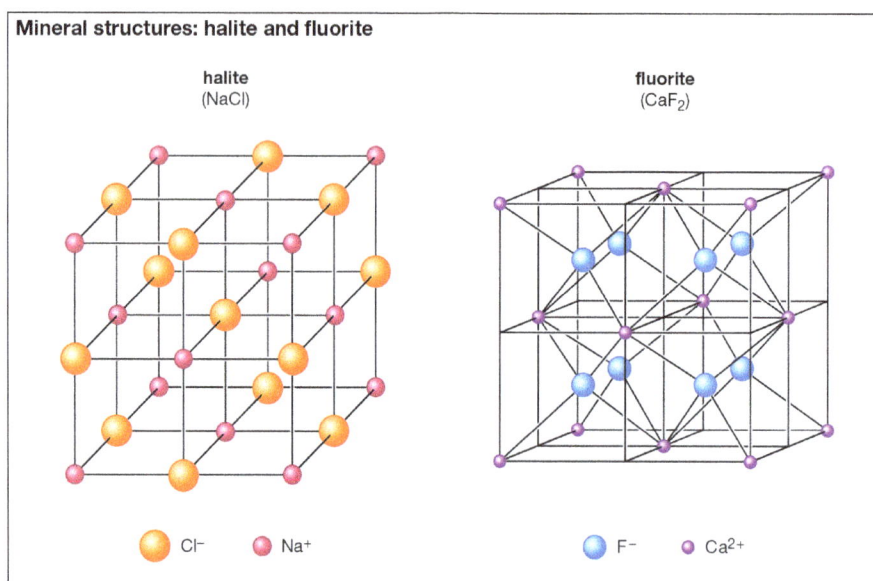

(A) The structure of halite, NaCl. (B) The structure of fluorite, CaF_2.

Several XZ_2 halides have the same structure as fluorite (CaF_2). In fluorite, calcium cations are positioned at the corners and face centres of cubic unit cells. (A unit cell is the smallest group of atoms, ions, or molecules from which the entire crystal structure can be generated by its repetition.) Each fluorine anion is in tetrahedral coordination with four calcium ions, while each calcium cation is in eightfold coordination with eight fluorine ions that form the corners of a cube around it. Uraninite (UO_2) and thorianite(ThO_2) are two examples of the several oxides that have a fluorite-type structure.

Carbonates

The carbonate minerals contain the anionic complex $(CO_3)^{2-}$, which is triangular in its coordination—i.e., with a carbon atom at the centre and an oxygen atom at each of the corners of an equilateral triangle. These anionic groups are strongly bonded individual units and do not share oxygen atoms with one another. The triangular carbonate groups are the basic building units of all carbonate minerals and are largely responsible for the properties particular to the class.

Carbonates are frequently identified using the effervescence test with acid. The reaction that results in the characteristic fizz, $2H^+ + CO^2/_3 \rightarrow H_2O + CO_2$, makes use of the fact that the carbon-oxygen bonds of the CO_3 groups are not quite as strong as the corresponding carbon-oxygen bonds in carbon dioxide.

The common anhydrous (water-free) carbonates are divided into three groups that differ in structure type: calcite, aragonite, and dolomite. The copper carbonates azurite and malachite are the only notable hydrous varieties.

The members of the calcite group share a common structure type. It can be considered as a derivative of the NaCl structure in which the carbonate (CO_3) group's substitute for the chlorine ions and calcium cations replace the sodium cations. As a result of the triangular shape of the CO_3 groups, the structure is rhombohedral instead of isometric as in NaCl. The CO_3 groups are in planes perpendicular to the threefold c-axis, and the calcium ions occupy alternate planes and are bonded to six oxygen atoms of the CO_3 groups.

Members of the calcite group exhibit perfect rhombohedral cleavage. The composition $CaCO_3$ most commonly occurs in two different polymorphs: rhombohedral calcite with calcium surrounded by six closest oxygen atoms and orthorhombic aragonite with calcium surrounded by nine closest oxygen atoms.

When CO_3 groups are combined with large divalent cations (generally with ionic radii greater than 1.0 Å), orthorhombic structures result. This is known as the aragonite structure type. Members of this group include those with large cations: $BaCO_3$, $SrCO_3$, and $PbCO_3$. Each cation is surrounded by nine closest oxygen atoms.

The aragonite group displays more limited solid solution than the calcite group. The type of cation present in aragonite minerals is largely responsible for the differences in physical properties among the members of the group. Specific gravity, for example, is roughly proportional to the atomic weight of the metal ions.

Dolomite $[CaMg(CO_3)_2]$, kutnohorite $[CaMn(CO_3)_2]$, and ankerite $[CaFe(CO_3)_2]$ are three isostructural members of the dolomite group. The dolomite structure can be considered as a calcite-type structure in which magnesium and calcium cations occupy the metal sites in alternate layers. The calcium (Ca^{2+}) and magnesium (Mg^{2+}) ions differ in size by 33 percent, and this produces cation ordering with the two cations occupying specific and separate levels in the structure. Dolomite has a calcium-to-magnesium ratio of approximately 1:1, which gives it a composition intermediate between $CaCO_3$ and $MgCO_3$.

Nitrates

The nitrates are characterized by their triangular $(NO_3)^-$ groups that resemble the $(CO_3)^{2-}$ groups of the carbonates, making the two mineral classes similar in structure. The nitrogen cation (N^{5+}) carries a high charge and is strongly polarizing like the carbon cation (C^{4+}) of the CO_3 group. A tightly knit triangular complex is created by the three nitrogen-oxygen bonds of the NO_3 group; these bonds are stronger than all others in the crystal. Because the nitrogen-oxygen bond has greater strength than the corresponding carbon-oxygen bond in carbonates, nitrates decompose less readily in the presence of acids.

Nitrate structures analogous to those of the calcite group result when NO_3 combines in a 1:1 ratio with monovalent cations whose radii can accommodate six closest oxygen neighbours. For example, nitratite ($NaNO_3$), also called soda nitre, and calcite exhibit the same structure, crystallography, and cleavage. The two minerals differ in that nitratite is softer and melts at a lower temperature owing to its lesser charge; also, sodium has a lower atomic weight than calcium, causing nitratite to have a lower specific gravity as well. Similarly, nitre (KNO_3), also known as saltpetre, is an analogue of aragonite. These are two examples of only seven known naturally occurring nitrates.

Borates

Minerals of the borate class contain boron-oxygen groups that can link together, in a phenomenon known as polymerization, to form chains, sheets, and isolated multiple groups. The silicon-oxygen (SiO_4) tetrahedrons of the silicates polymerize in a manner similar to the $(BO_3)^{3-}$ triangular groups of the borates. A single oxygen atom is shared between two boron cations (B^{3+}), thereby linking the BO_3 groups into extended units such as double triangles, triple rings, sheets, and chains. The oxygen atom is able to accommodate two boron atoms because the small boron cation has a bond strength to each oxygen that is exactly one-half the bond energy of the oxygen ion.

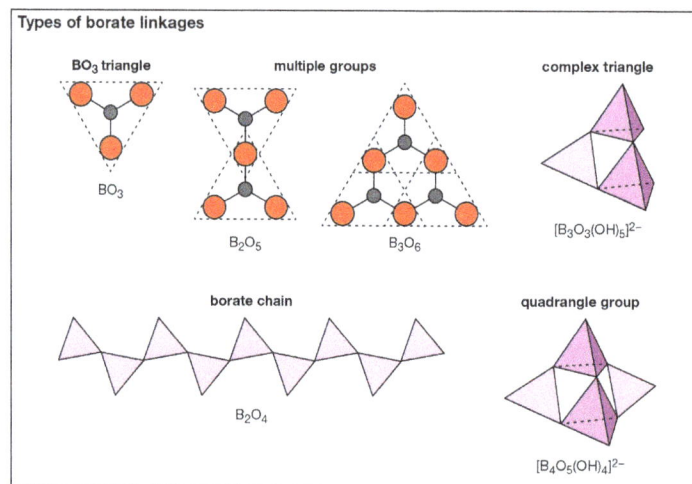

Various possible linkages of (A) BO_3 triangles to form (B,C) multiple groups and (D) chains in borates. Complex (E) triangle and (F) quadrangle groups are also shown. The group depicted in (F) occurs in borax.

Although boron is usually found in triangular coordination with three oxygens, it also occurs in fourfold coordination in tetrahedral groups. In addition, boron may exist as part of complex anionic groups such as $[B_3O_3(OH)_3]^{2-}$, consisting of one triangle and two tetrahedrons. Complex infinite chains of tetrahedrons and triangles are found in the structure of colemanite $[CaB_3O_4(OH)_3 \cdot H_2O]$; a complex ion composed of two tetrahedrons and two triangles, $[B_4O_5(OH)_4]^{2-}$, is present in borax $[Na_2B_4O_5(OH)_4 \cdot 8H_2O]$.

Sulfates

This class is composed of a large number of minerals, but relatively few are common. All contain anionic $(SO_4)^{2-}$ groups in their structures. These anionic complexes are formed through the tight bonding of a central S^{6+} ion to four neighbouring oxygen atoms in a tetrahedral arrangement around the sulfur. This closely knit group is incapable of sharing any of its oxygen atoms with other SO_4 groups; as such, the tetrahedrons occur as individual, unlinked groups in sulfate mineral structures.

Common sulfates	
Barite group	
Barite	$BaSO_4$
Celestite	$SrSO_4$
Anglesite	$PbSO_4$
Anhydrite	$CaSO_4$
Gypsum	$CaSO_4 \cdot 2H_2O$

Members of the barite group constitute the most important and common anhydrous sulfates. They have orthorhombic symmetry with large divalent cations bonded to the sulfate ion. In barite $(BaSO_4)$, each barium ion is surrounded by 12 closest oxygen ions belonging to seven distinct SO_4 groups. Anhydrite $(CaSO_4)$ exhibits a structure very different from that of barite since the ionic radius of Ca^{2+} is considerably smaller than Ba^{2+}. Each calcium cation can only fit eight oxygen atoms around it from neighbouring SO_4 groups. Gypsum $(CaSO_4 \cdot 2H_2O)$ is the most important and abundant hydrous sulfate.

Phosphates

Although this mineral class is large (with almost 700 known species), most of its members are quite rare. Apatite $[Ca_5(PO_4)_3(F, Cl, OH)]$, however, is one of the most important and abundant phosphates. The members of this group are characterized by tetrahedral anionic $(PO_4)^{3-}$ complexes, which are analogous to the $(SO_4)^{2-}$ groups of the sulfates. The phosphorus ion, with a valence of positive five, is only slightly larger than the sulfur ion, which carries a positive six charge. Arsenates and vanadates are similar to phosphates.

Silicates

The silicates, owing to their abundance on Earth, constitute the most important mineral class. Approximately 25 percent of all known minerals and 40 percent of the most common ones are silicates; the igneous rocksthat make up more than 90 percent of Earth's crust are composed of virtually all silicates.

The fundamental unit in all silicate structures is the silicon-oxygen $(SiO_4)^{4-}$ tetrahedron. It is composed of a central silicon cation (Si^{4+}) bonded to four oxygen atoms that are located at the corners of a regular tetrahedron. The terrestrial crust is held together by the strong silicon-oxygen bonds of these tetrahedrons. Approximately 50 percent ionic and 50 percent covalent, the bonds develop from the attraction of oppositely charged ions as well as the sharing of their electrons.

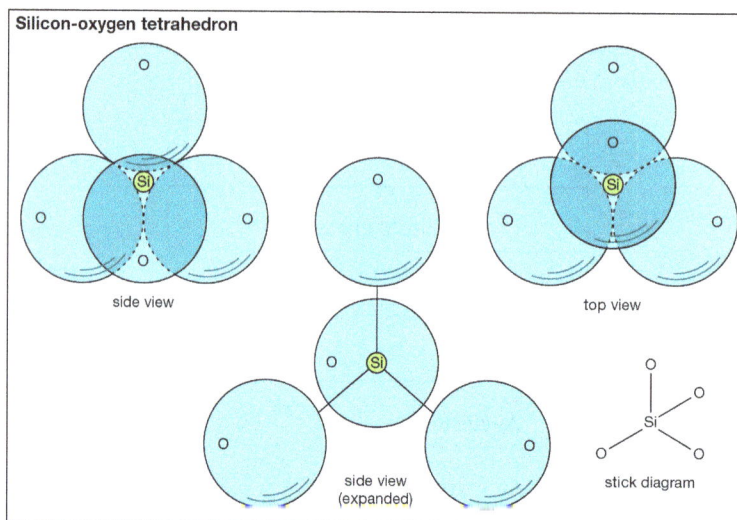

Two views of a closest-packed representation of the silicon-oxygen tetrahedron.

The positive charge (+4) of each silicon cation is satisfied by its four bonds to oxygen atoms. Each oxygen ion (O^{2-}), however, contributes only one-half of its totals bonding energy to a silicon-oxygen bond, so it is capable of also bonding to the silicon cation of another tetrahedron. The SiO_4 tetrahedrons thereby become linked by shared oxygen atoms; this is referred to as polymerization. The degree and manner of polymerization are the bases for the variety present in silicate structures.

The silicates can be divided into groups according to structural configuration, which arises from the sharing of one, two, three, or all oxygen ions of a tetrahedron. Nesosilicates have isolated groups of SiO_4, while sorosilicates contain pairs of SiO_4 tetrahedrons linked into Si_2O_7 groups. Ring silicates, also known as cyclosilicates, are closed, ringlike silicates; the sixfold variety has composition Si_6O_{18}. Silicates that are composed of infinite chains of tetrahedrons are called inosilicates; single chains have a unit composition of SiO_3 or Si_2O_6, whereas double chains contain a silicon to oxygen ratio of 4:11. Phyllosilicates, or sheet silicates, are formed when three oxygen atoms are shared with adjoining tetrahedrons. The resulting infinite flat sheets have unit composition Si_2O_5. In structures where tetrahedrons share all their oxygen ions, an infinite three-dimensional network is created with an SiO_2 unit composition. Minerals of this type are called framework silicates or tectosilicates.

As a major constituent of Earth's crust, aluminum follows only oxygen and silicon in importance. The radius of aluminum, slightly larger than that of silicon, lays close to the upper bound for allowable fourfold coordination in crystals. As a result, aluminum can be surrounded with four oxygen atoms arranged tetrahedrally, but it can also occur in sixfold coordination with oxygen. The ability to maintain two roles within the silicate structure makes aluminum a unique constituent of these minerals. The tetrahedral AlO_4 groups are approximately equal in size to SiO_4 groups and therefore can become incorporated into the silicate polymerization scheme. Aluminum in sixfold coordination may form ionic bonds with the SiO_4 tetrahedrons. Thus, aluminum may occupy tetrahedral sites as a replacement for silicon and octahedral sites in solid solution with elements such as magnesium and ferrous iron.

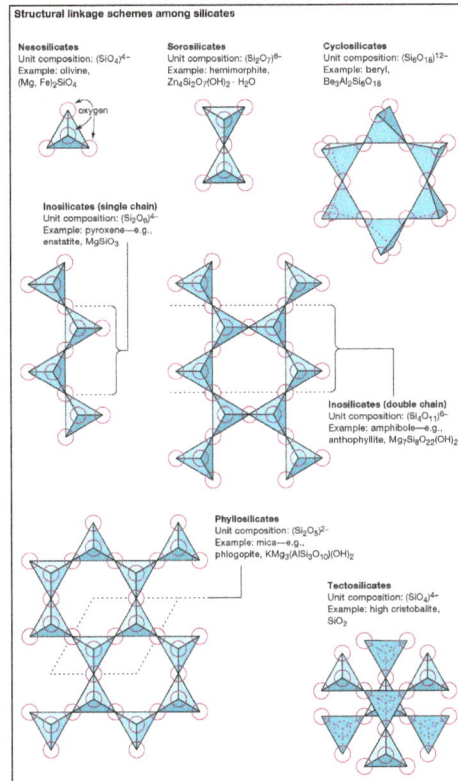

Various structural linkage schemes in silicates.

Several ions may be present in silicate structures in octahedral coordination with oxygen: Mg^{2+}, Fe^{2+}, Fe^{3+}, Mn^{2+}, Al^{3+}, and Ti^{4+}. All cations have approximately the same dimensions and thus are found in equivalent atomic sites, even though their charges range from positive two to positive four. Solid solution involving ions of different charge is accomplished through coupled substitutions, thereby maintaining neutrality of the structures.

Nesosilicates

The silicon-oxygen tetrahedrons of the nesosilicates are not polymerized they are linked to one another only by ionic bonds of the interstitial cations. As a result of the isolation of the tetrahedral groups, the crystal habits of these minerals are typically equi-dimensional so that prominent cleavage directions are not present. The size and charge of the interstitial cations largely determine the structural form of the nesosilicates. The relatively high specific gravity and hardness that are characteristic of this group arise from the dense packing of the atoms within the structure. Substitution of aluminum for silicon is normally quite low.

Sorosilicates

These minerals contain sets of two SiO_4 tetrahedrons joined by one shared apical oxygen. A silicon-to-oxygen ratio of 2:7 is consequently present in their structures. More than 70 minerals belong to the sorosilicate group, although most are rare. Only the members of the epidote group and vesuvianite are common. Both independent $(SiO_4)^{4-}$ and double $(Si_2O_7)^{6-}$ groups are incorporated into the epidote structure, as is reflected in its formula: $Ca_2(Al, Fe)Al_2O(SiO_4)(Si_2O_7)(OH)$.

Cyclosilicates

Silicon-oxygen tetrahedrons are linked into rings in cyclosilicate structures, which have an overall Si:O ratio of 1:3. There are three closed cyclic configurations with the following formulas: Si_3O_9, Si_4O_{12}, and Si_6O_{18}. The rare titanosilicate benitoite $(BaTiSi_3O_9)$ is the only mineral that is built with the simple Si_3O_9 ring. Axinite $[(Ca, Fe, Mn)_3Al_2(BO_3)(Si_4O_{12})(OH)]$ contains Si_4O_{12} rings, along with BO_3 triangles and OH groups. The two common and important cyclosilicates, beryl $(Be_3Al_2Si_6O_{18})$ and tourmaline (which has an extremely complex formula), are based on the Si_6O_{18} ring.

Inosilicates

This class is characterized by its one-dimensional chains and bands created by the linkage of SiO_4 tetrahedrons. Single chains may be formed by the sharing of two oxygen atoms from each tetrahedron, resulting in a structure with an Si:O ratio of 1:3. Two such chains that are aligned side by side with alternate tetrahedrons sharing an additional oxygen atom form bands of double chains. These structures have an Si:O ratio of 4:11. There are a number of silicate minerals, pyroxenoids, which have a similar Si:O ratio as pyroxene, but with structures that are not identical as the chains of silicon tetrahedra do not infinitely repeat. Two significant rock-forming mineral families display these structure types: the single-chain pyroxenes and the double-chain amphiboles.

Inosilicates: Common pyroxenes and amphiboles	
Pyroxenes	
Enstatite-orthoferrosilite series	
Enstatite	$MgSiO_3$
Orthoferrosilite	$FeSiO_3$
Diopside-hedenbergite series	
Diopside	$CaMgSi_2O_6$
Hedenbergite	$CaFeSi_2O_6$
Augite	$(Ca, Na)(Fe, Mg, Al)(Al, Si)_2O_6$
Sodium pyroxene group	
Jadeite	$NaAlSi_2O_6$
Acmite	$NaFe^{3+}Si_2O_6$
Amphiboles	
Anthophyllite	$(Mg, Fe)_7Si_8O_{22}(OH)_2$
Cummingtonite series	
Cummingtonite	$Fe_2Mg_5Si_8O_{22}(OH)_2$
Grunerite	$Fe_7Si_8O_{22}(OH)_2$
Tremolite series	

Tremolite	$Ca_2Mg_5Si_8O_{22}(OH)_2$
Actinolite	$Ca_2(Mg, Fe)_5Si_8O_{22}(OH)_2$
Hornblende	$(Ca, Na)_2(Mg, Fe, Al)_5(Si, Al)_8O_{22}(OH)_2$
Sodic amphibole group	
Glaucophane	$Na_2Mg_3Al_2Si_8O_{22}(OH)_2$
Riebeckite	$Na_2Fe_3^{2+}Fe_2^{3+}Si_8O_{22}(OH)_2$

The amphiboles and pyroxenes share the same cations and have many similar crystallographic, chemical, and physical properties: the colour, lustre, and hardness of analogous species are alike. A distinguishing factor between the two groups, the presence of the hydroxyl radical in the amphiboles, generally gives the double-chain members lower specific gravities and refractive indices than their single-chain analogues. Their crystal habits also are different: amphiboles exhibit needlelike or fibrous crystals, while pyroxenes take the form of stubby prisms. In addition, the different chain structures of the two groups result in different cleavage angles.

Pyroxenes occur in high-temperature igneous and metamorphic rocks. They crystallize at higher temperatures than their amphibole counterparts. A pyroxene formed early in the cooling of an igneous melt or in a metamorphic fluid may later combine with water at a lower temperature to form amphibole.

Phyllosilicates

These minerals display a two-dimensional framework of infinite sheets of SiO_4 tetrahedrons. An Si:O ratio of 2:5 results from the sharing of three oxygen atoms in each tetrahedron. Sixfold symmetry is exhibited in undistorted sheets. The silicate sheet framework is largely responsible for the following properties of the phyllosilicates: platy or flaky habit, single pronounced cleavage, low specific gravity, softness, and possible flexibility and elasticity of cleavage layers. Most minerals of this group contain hydroxyls positioned in the middle of the sixfold ring of tetrahedrons.

Many soil constituents, produced through rock weathering, possess a sheet structure. Phyllosilicate properties contribute greatly to the ability of soils to release and retain plant food, to reserve water from wet to dry seasons, and to accommodate organisms and atmospheric gases.

Tectosilicates

Almost 75 percent of Earth's crust is composed of minerals with the three-dimensional framework of the tectosilicates. All oxygen atoms of the SiO_4 tetrahedrons of members of this class are shared with nearby tetrahedrons, creating a strongly bound structure with an Si:O ratio of 1:2. Other than the zeolite group, which can accommodate water owing to the open nature of its structure, all members listed in the table are anhydrous.

Mineral Associations and Phase Equilibrium

A phase is a homogeneous substance that has a fixed composition and uniform chemical and physical properties. Only a mineral that displays no solid solution may therefore be considered a phase. Quartz (SiO_2), for example, is a low-temperature phase in the $Si-O_2(SiO_2)$ system, and kyanite(Al_2SiO_5) is a high-pressure phase in the $Al_2O_3SiO_2(Al_2SiO_5)$ system. The term phase region is used when a mineral exhibits compositional variation, as in the solid-solution series between forsterite and fayalite. A phase may exist as a solid, liquid, or gas: H_2O, for example, occurs in the form of ice (solid), water (liquid), and steam (gas).

Equilibrium refers to the stable coexistence of two or more phases and is established relative to time. If two phases in a mixture of water and ice coexist so that the amount of each is fixed indefinitely, they are said to be in equilibrium. The minerals of some rocks have existed together since their formation for periods of several million years, yet one cannot always ascertain if these rock constituents are in equilibrium or are still undergoing changes.

A determining factor of the equilibrium state of minerals is the presence (or absence) of a re-action rim, which is a region separating two or more minerals and consisting of the products of a reaction between them. The absence of any observable reaction rims between minerals that physically touch each other suggests that they were in equilibrium at the time when the rock formed. Additional chemical data regarding elemental distribution between the minerals is necessary to verify this assumption. In contrast, the presence of megascopically or microscopically visible rims indicates that some minerals were not in equilibrium. Garnet, for example, may react with coexisting biotite to produce a chlorite rim between them, revealing that the two minerals were not always in equilibrium. An experimental petrologist must assign some period of time after which the absence of further changes between phases will indicate that equilibrium has been reached. The time period is variable, depending on the speed of the reactions involved and in part on the patience of the investigator; it may range from a few hours to several years.

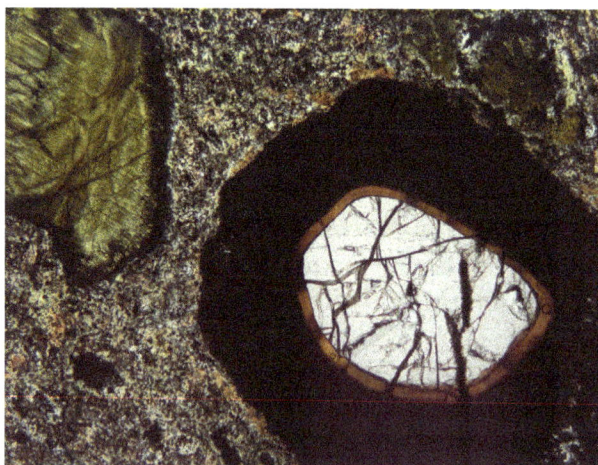

Reaction rim: A brown reaction rim encircling a garnet xenocryst. A determining factor of the equilibrium state of minerals is the presence (or absence) of a reaction rim, a region separating two or more minerals and consisting of the products of a reaction between them.

Components are the minimum number of independent chemical species that are necessary to de-scribe the compositions of all the phases present in a system. The compound H_2O is generally used as the sole component defining the H_2O system, although H_2 and O_2 define the chemical system as well. In examinations of the stability fields of $MgSiO_3$ (enstatite), $MgSiO_3$ is normally used as the com-ponent rather than the three elements, Mg, Si, and O, or the two oxides, MgO and SiO_2. The three components generally used in the pyroxene system $CaO-MgO-FeO-SiO_2$ are $CaSiO_3-MgSiO_3-FeSiO_3$.

Assemblage and the Phase Rule

In the early stages of the study of a rock, the constituent minerals of the rock must be identified. Orthoclase, albite, quartz, and biotite may be found in an igneous granite. By examining the granite's texture, one may conclude that the four minerals crystallized at approximately the same elevated temperature and that orthoclase-albite-quartz-biotite is its mineral assemblage. The term assemblage is frequently applied to all minerals included in a rock but more appropriately should be used for those minerals that are in equilibrium (and are known more specifically as the equilibrium assemblage). The granite may display surficial cavities that are lined by several clay minerals and limonite (a hydrous iron oxide). The original high-temperature granite was altered to form the low-temperature clay minerals and limonite; there are consequently two distinct assemblages present in the rock: the high-temperature orthoclase-albite-quartz-biotite assemblage and the low-temperature assemblage of clay minerals and limonite.

Metamorphic rocks also may contain separate assemblages. A shale that at low temperatures was composed of a sericite-kaolin-dolomite-quartz-feldspar assemblage can become metamorphosed at higher temperatures to produce a garnet-sillimanite-biotite-feldspar assemblage.

An assemblage thus consists of minerals that formed under the same or quite comparable conditions of pressure and temperature. In practice, minerals that physically touch one another with no reaction rims or alteration products are included in the assemblage. It is likely that the minerals satisfying these conditions are in equilibrium, but additional chemical tests are commonly necessary to define the equilibrium assemblage without ambiguity.

Phase systems are governed by a phase rule, which defines the number of minerals that may coexist in equilibrium: $F = C - P + 2$, where F is the variance, or number of degrees of freedom, C is the number of independent components, and P is the number of phases. Applying this rule to a three-phase, three-component system, F is 2. This indicates that two parameters—e.g., pressure and temperature—may be varied independently of one another without altering the number of phases.

Phase Diagrams

Phase (or stability) diagrams are used to illustrate the conditions under which certain minerals are stable. They are graphs that show the limiting conditions for solid, liquid, and gaseous phases of a single substance or of a mixture of substances while undergoing changes in pressure and temperature or in some other combination of variables. The following are examples of phase diagrams employed in the study of igneous, metamorphic, and sedimentary rocks.

Use in Igneous Petrology

In the field of igneous petrology, the researcher commonly employs a phase equilibrium approach to compare the mineral assemblages found in naturally occurring and synthetic rocks. Much can be learned from studying the melting of an igneous rock and the reverse process, the crystallization of minerals from a melt (liquid phase). Graphic representations of systems with a liquid phase are called liquidus diagrams. The dashed contours of a liquidus diagram, called isotherms, represent temperatures at which a mineral melts. They define what is known as a liquidus surface. As temperatures decrease, the minerals will crystallize in the manner defined by the arrows on the boundaries separating the different mineral phases. A careful study of the crystalline products formed upon the cooling of melts of specific compositions allows the igneous petrologist to compare such results with minerals observed in natural igneous rocks.

Use in Metamorphic Petrology

Pressure-temperature (P-T) phase diagrams are applied in the study of the conditions under which metamorphic rocks originate. They illustrate the equilibrium relationships among various mineral phases in terms of pressure and temperature. The minerals that are separated by a reaction curve may exist in equilibrium at the conditions occurring along lines present within the diagram. For example, the reaction curves for Al_2SiO_5 and for muscovite + quartz $\leftrightarrow$ potassium-feldspar + sillimanite + H_{2O} are significant in metamorphic rocks that have a high aluminum oxide (Al_2O_3) content as compared to other components (e.g., calcium oxide [CaO], magnesium oxide [MgO], and ferrous oxide [FeO]). Shales enriched in clay minerals contain a rather large amount of aluminum oxide, and during metamorphism of the shale mineral reactions and recrystallization occur. In their metamorphic form, shales appear as pelitic schists, and these may include significant amounts of sillimanite, muscovite, and quartz. Such schist may have equilibrated under a certain set of pressure and temperature conditions.

Theoretical calculations are combined with experimental observations to arrive at phase diagrams. In laboratory experiments conducted on the three polymorphs of Al_2SiO_5, chemicals of high purity are most often used as the starting materials, but extremely pure minerals may be substituted. A specimen of gem-grade kyanite that does not contain any inclusions may be reached at high temperatures to form sillimanite. The placement of the reaction curve between the kyanite and sillimanite fields is determined by the first instance of sillimanite formation from kyanite and also the initial stage of the reverse reaction. X-ray powder diffraction and optical microscopic techniques are employed to estimate the conditions under which these reactions commence, but the experimental methods are subject to some degree of uncertainty. Therefore, the reaction curves, although commonly drawn as narrow lines, may actually represent wider reaction zones. Also, the naturally occurring system is more complex than the simplified version devised in the laboratory, and so difficulty arises when attempting to relate the two. A P-T diagram serves mainly as a tool in evaluating the conditions of metamorphism, such as pressure and temperature.

Use in Sedimentary Petrology

Phase diagrams can also be helpful in the assessment of physical and chemical conditions that prevailed during the deposition of a chemical sedimentary sequence. Atmospheric conditions are characterized by low temperatures and pressures, and under such conditions stability fields of minerals can often conveniently be expressed in terms of Eh (oxidationpotential) and pH (the negative logarithm of the hydrogen ion concentration [H^+]; a pH of 0–7 indicates acidity, a pH of 7–14 indicates basicity, and neutral solutions have a pH of 7).

Stability relations for some iron oxides and iron sulfides are often presented at atmospheric conditions, 25 °C (77 °F) and one standard atmospherepressure. (One standard atmosphere of pressure equals 760 millimetres, or 29.92 inches, of mercury.) A high Eh value corresponds to a compoundstable under oxidizing conditions, such as hematite, while a low Eh value indicates a mineral that occurs in reducing environments, such as magnetite. Pyrite and pyrrhotite, two sulfide minerals, occur at low Eh values and at pH values of 4–9. Lines separating the fields of an Eh-pH diagram represent conditions under which the two minerals may exist in equilibrium. Hematite and magnetite, for example, are often found together in iron-bearing sediments. Eh-pH diagrams are valuable in providing information regarding the chemical and physical environments that existed during atmospheric weathering and during chemical sedimentation and diagenesis of sediments deposited by water at temperatures of 25 to about 100 °C (77 to about 212 °F) and at a pressure of approximately one atmosphere. The coexistence of hematite and magnetite common in Precambrian iron-bearing rocks (those formed from 4.6 billion to 541 million years ago) may enable investigators to estimate variables such as Eh and pH that prevailed in the original ancient sedimentary basin.

Petrology

Petrology is a field of geology that focuses on the study of rocks and the conditions under which they are formed. It utilizes the classical fields of mineralogy, petrography, and chemical analyses to describe the structure and composition of rocks. In addition, modern petrologists include the principles of geochemistry and geophysics to better understand the origins of rocks. There are three branches of petrology, corresponding to the three main types of rocks: igneous, metamorphic, and sedimentary.

A sample of igneous rock. The light-colored tracks show the direction of lava flow.

The study of rocks provides us with important information about the nature of the Earth's crust and mantle. In addition, it enables us to gain a sense of the Earth's history, including tectonic processes that occurred over the long course of geological time. On a practical level, the field of petrology helps us gain an understanding of many of the raw materials we rely on for our sustenance and technological development.

A sample of quartzite, a form of metamorphic rock.

Branches of Petrology

- Igneous petrology focuses on the composition and texture of igneous rocks (rocks such as granite or basalt which have crystallized from molten rock or magma). Igneous rocks include volcanic and plutonic rocks.

- Sedimentary petrology focuses on the composition and texture of sedimentary rocks (rocks such as sandstone, shale, or limestone which consist of pieces or particles derived from other rocks or biological or chemical deposits, and are usually bound together in a matrix of finer material).

- Metamorphic petrology focuses on the composition and texture of metamorphic rocks (rocks such as slate, marble, gneiss, or schist which started out as sedimentary or igneous rocks but which have undergone chemical, mineralogical or textural changes due to extremes of pressure, temperature or both).

- Experimental petrology employs high-pressure, high-temperature apparatus to investigate the geochemistry and phase relations of natural or synthetic materials at elevated pressures and temperatures. Experiments are particularly useful for investigating rocks of the lower crust and upper mantle that rarely survive the journey to the surface in pristine condition. The work of experimental petrologists has laid a foundation on which modern understanding of igneous and metamorphic processes has been built.

Two types of sedimentary rock, limey shale overlaid by
limestone, observed at Cumberland Plateau, Tennessee.

Significance of Studying Rocks

The study of rocks is important for several reasons:

- Their minerals and global chemistry provide information about the composition of the Earth's crust and mantle.

- Their ages can be calculated by various methods of radiometric dating, and a time sequence of geological events can be put together.

- Their features are usually characteristic of a specific tectonic environment, allowing scientists to reconstitute tectonic processes.

- Many rocks host important ores that provide valuable raw materials that we rely on for our sustenance and technological development.

Rocks

In Geology, Rock is naturally occurring and coherent aggregate of one or more minerals. Such aggregates constitute the basic unit of which the solid Earth is comprised and typically form recognizable and mappable volumes. Rocks are commonly divided into three major classes according to the processes that resulted in their formation. These classes are (1) igneous rocks, which have solidified from molten material called magma; (2) sedimentary rocks, those consisting of fragments derived from preexisting rocks or of materials precipitated from solutions; and (3) metamorphic rocks, which have been derived from either igneous or sedimentary rocks under conditions that caused changes in mineralogical composition, texture, and internal structure. These three classes, in turn, are subdivided into numerous groups and types on the basis of various factors, the most important of which are chemical, mineralogical, and textural attributes.

Texture

The texture of a rock is the size, shape, and arrangement of the grains (for sedimentary rocks) or crystals (for igneous and metamorphic rocks). Also of importance are the rock's extent of homogeneity (*i.e.,* uniformity of composition throughout) and the degree of isotropy. The latter is the extent to which the bulk structure and composition are the same in all directions in the rock.

obsidian   porphyry   calico, or laminated sandstone   coquina, or shell limestone

breccia   banded gneiss   talc schist   serpentine

In figure, Rocks have many different textures. Layered sandstone produces a gritty texture, whereas coquina may be rough with cemented shells occasionally producing a sharp edge. Likewise, breccia, which contains pieces of other rocks that have been cemented together, and porphyry, which contains interlocking mineral crystals, tend to be rough. In contrast, obsidian tends to have a smooth glassy feel, whereas serpentine may feel platy or fibrous, and talc schist often feels greasy. On the other hand, the texture of gneiss is often described by its distinct banding.

Analysis of texture can yield information about the rock's source material, conditions and environment of deposition (for sedimentary rock) or crystallization and recrystallization (for igneous and metamorphic rock, respectively), and subsequent geologic history and change.

Classification by Grain or Crystal Size

The common textural terms used for rock types with respect to the size of the grains or crystals, are given in the Table. The particle-size categories are derived from the Udden-Wentworth scale developed for sediment. For igneous and metamorphic rocks, the terms are generally used as modifiers—*e.g.*, medium-grained granite. Aphanitic is a descriptive term for small crystals, and phaneritic for larger ones. Very coarse crystals (those larger than 3 centimetres, or 1.2 inches) are termed pegmatitic.

Table: Common textual terms for rocks.

size (in millimetres)	igneous and metamorphic		sedimentary		pyroclastic	
			sediment	rock	sediment	rock
256		very coarse (pegmatitic)	boulder			
128			cobble		block bomb	breccia
64						
32				conglomerate		
16	phaneritic	coarse	pebble			
8					lapilli cinder	lapilli
4		medium	granule			
2						
1				coarse sandstone	coarse ash	coarse tuff
1/2		fine	sand			
1/4				fine sandstone		
1/8						
1/16						
1/32	aphanitic		silt	siltstone	fine ash	fine tuff
1/64						
1/128		dense				
1/256						
			clay	shale		

*Diagonal lines in the table reflect the variability in size limits of certain grades resulting from the use of different values by different authors.

For sedimentary rocks, the broad categories of sediment size are coarse (greater than 2 millimetres, or 0.08 inch), medium (between 2 and $^1/_{16}$ millimetres), and fine (under $^1/_{16}$ millimetre). The latter includes silt and clay, which both have a size indistinguishable by the human eye and are also termed dust. Most shales (the lithified version of clay) contain some silt. Pyroclastic rocks are those formed from clastic material ejected from volcanoes. Blocks are fragments broken from solid rock, while bombs are molten when ejected.

Porosity

The term rock refers to the bulk volume of the material, including the grains or crystals as well as the contained void space. The volumetric portion of bulk rock that is not occupied by grains, crystals, or natural cementing material is termed porosity. That is to say, porosity is the ratio of void volume to the bulk volume (grains plus void space). This void space consists of pore space between grains or crystals, in addition to crack space. In sedimentary rocks, the amount of pore space depends on the degree of compaction of the sediment (with compaction generally increasing with depth of burial), on the packing arrangement and shape of grains, on the amount of cementation, and on the degree of sorting. Typical cements are siliceous, calcareous or carbonate, or iron-bearing minerals.

Sorting is the tendency of sedimentary rocks to have grains that are similarly sized—*i.e.*, to have a narrow range of sizes. Poorly sorted sediment displays a wide range of grain sizes and hence has decreased porosity. Well-sorted indicates a grain size distribution that is fairly uniform. Depending on the type of close-packing of the grains, porosity can be substantial. It should be noted that in engineering usage—*e.g.*, geotechnical or civil engineering—the terminology is phrased oppositely and is referred to as grading. Well-graded sediment is a (geologically) poorly sorted one, and poorly graded sediment is a well-sorted one.

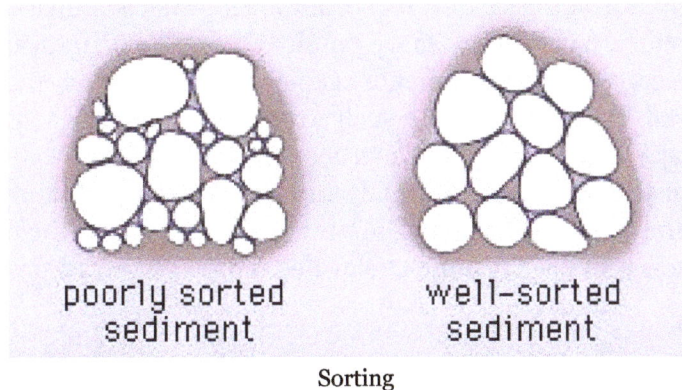

Sorting

Total porosity encompasses all the void space, including those pores that are interconnected to the surface of the sample as well as those that are sealed off by natural cement or other obstructions. Thus the total porosity (φ_T) is:

$$\varphi_T = 100\left(1 - \frac{Vol_G}{Vol_B}\right)\%$$

where Vol_G is the volume of grains (and cement, if any) and Vol_B is the total bulk volume. Alternatively, one can calculate φ_T from the measured densities of the bulk rock and of the (mono) mineralic constituent. Thus,

$$\varphi_T = 100\left(1 - \frac{p_B}{p_G}\right)\%.$$

where ρ_B is the density of the bulk rock and ρ_G is the density of the grains (*i.e.*, the mineral, if the composition is monomineralogic and homogeneous). For example, if a sandstone has a ρ_B of 2.38 grams per cubic centimetre (g/cm³) and is composed of quartz (SiO_2) grains having ρG of 2.65 g/cm³, the total porosity is:

$$\varphi_T = 100\left(1 - \frac{2.38}{2.65}\right) = 10.2\%.$$

Apparent (effective, or net) porosity is the proportion of void space that excludes the sealed-off pores. It thus measures the pore volume that is effectively interconnected and accessible to the surface of the sample, which is important when considering the storage and movement of subsurface fluids such as petroleum, groundwater, or contaminated fluids.

Physical Properties

Physical properties of rocks are of interest and utility in many fields of work, including geology, petrophysics, geophysics, materials science, geochemistry, and geotechnical engineering. The scale of investigation ranges from the molecular and crystalline up to terrestrial studies of the Earth and other planetary bodies. Geologists are interested in the radioactive age dating of rocks to reconstruct the origin of mineral deposits; seismologists formulate prospective earthquake predictions using premonitory physical or chemical changes; crystallographers study the synthesis of minerals with special optical or physical properties; exploration geophysicists investigate the variation of physical properties of subsurface rocks to make possible detection of natural resources such as oil and gas, geothermal energy, and ores of metals; geotechnical engineers examine the nature and behaviour of the materials on, in, or of which such structures as buildings, dams, tunnels, bridges, and underground storage vaults are to be constructed; solid-state physicists study the magnetic, electrical, and mechanical properties of materials for electronic devices, computer components, or high-performance ceramics; and petroleum reservoir engineers analyze the response measured on well logs or in the processes of deep drilling at elevated temperature and pressure.

Since rocks are aggregates of mineral grains or crystals, their properties are determined in large part by the properties of their various constituent minerals. In a rock these general properties are determined by averaging the relative properties and sometimes orientations of the various grains or crystals. As a result, some properties that are anisotropic (*i.e.*, differ with direction) on a submicroscopic or crystalline scale are fairly isotropic for a large bulk volume of the rock. Many properties are also dependent on grain or crystal size, shape, and packing arrangement, the amount and distribution of void space, the presence of natural cements in sedimentary rocks, the temperature and pressure, and the type and amount of contained fluids (*e.g.*, water, petroleum, gases). Because many rocks exhibit a considerable range in these factors, the assignment of representative values for a particular property is often done using a statistical variation.

Some properties can vary considerably, depending on whether measured in situ (in place in the subsurface) or in the laboratory under simulated conditions. Electrical resistivity, for example, is highly dependent on the fluid content of the rock in situ and the temperature condition at the particular depth.

Density

Density varies significantly among different rock types because of differences in mineralogy and porosity. Knowledge of the distribution of underground rock densities can assist in interpreting subsurface geologic structure and rock type.

In strict usage, density is defined as the mass of a substance per unit volume; however, in common usage, it is taken to be the weight in air of a unit volume of a sample at a specific temperature. Weight is the force that gravitation exerts on a body (and thus varies with location), whereas mass (a measure of the matter in a body) is a fundamental property and is constant regardless of location. In routine density measurements of rocks, the sample weights are considered to be equivalent to their masses, because the discrepancy between weight and mass would result in less error on the computed density than the experimental errors introduced in the measurement of volume. Thus, density is often determined using weight rather than mass. Density should properly be reported in kilograms per cubic metre (kg/m³), but is still often given in grams per cubic centimetre (g/cm³).

Another property closely related to density is specific gravity. It is defined as the ratio of the weight or mass in air of a unit volume of material at a stated temperature to the weight or mass in air of a unit volume of distilled water at the same temperature. Specific gravity is dimensionless (*i.e.*, has no units).

The bulk density of a rock is $\rho_B = W_G/V_B$, where WG is the weight of grains (sedimentary rocks) or crystals (igneous and metamorphic rocks) and natural cements, if any, and V_B is the total volume of the grains or crystals plus the void (pore) space. The density can be dry if the pore space is empty, or it can be saturated if the pores are filled with fluid (*e.g.*, water), which is more typical of the subsurface (in situ) situation. If there is more fluid present,

$$\text{(saturated) } \rho_B = \frac{W_G + W_{fl}}{V_B},$$

where W_{fl} is the weight of pore fluid. In terms of total porosity, saturated density is:

$$\text{(saturated) } \rho_B = \rho_G(1 - \varphi_T) + p_{fl\varphi_T},$$

And thus,

$$\varphi_T = \frac{\rho_G - \rho_B}{\rho_G - \rho_{fl}},$$

where ρ_{fl} is the density of the pore fluid. Density measurements for a given specimen involve the determination of any two of the following quantities: pore volume, bulk volume, or grain volume, along with the weight.

A useful way to assess the density of rocks is to make a histogram plot of the statistical range of a set of data. The representative value and its variation can be expressed as follows: (1) mean, the average value, (2) mode, the most common value (*i.e.*, the peak of the distribution curve), (3) median, the value of the middle sample of the data set (*i.e.*, the value at which half of the samples are below and half are above), and (4) standard deviation, a statistical measure of the spread of the data (plus and minus one standard deviation from the mean value includes about two-thirds of the data).

Mechanical Properties

Stress and Strain

When a stress σ (force per unit area) is applied to a material such as rock, the material experiences a change in dimension, volume, or shape. This change, or deformation, is called strain (ε). Stresses can be axial—*e.g.*, directional tension or simple compression—or shear (tangential), or all-sided (*e.g.*, hydrostatic compression). The terms stress and pressure are sometimes used interchangeably, but often stress refers to directional stress or shear stress and pressure (*P*) refers to hydrostatic compression. For small stresses, the strain is elastic (recoverable when the stress is removed and linearly proportional to the applied stress). For larger stresses and other conditions, the strain can be inelastic, or permanent.

Elastic Constants

In elastic deformation, there are various constants that relate the magnitude of the strain response to the applied stress. These elastic constants include the following:

1. Oung's modulus (E) is the ratio of the applied stress to the fractional extension (or shortening) of the sample length parallel to the tension (or compression). The strain is the linear change in dimension divided by the original length.

2. Shear modulus (μ) is the ratio of the applied stress to the distortion (rotation) of a plane originally perpendicular to the applied shear stress; it is also termed the modulus of rigidity.

3. Bulk modulus (k) is the ratio of the confining pressure to the fractional reduction of volume in response to the applied hydrostatic pressure. The volume strain is the change in volume of the sample divided by the original volume. Bulk modulus is also termed the modulus of incompressibility.

4. Poisson's ratio (σ_p) is the ratio of lateral strain (perpendicular to an applied stress) to the longitudinal strain (parallel to applied stress).

For elastic and isotropic materials, the elastic constants are interrelated. For example,

$$\sigma_p = \frac{E}{2\mu} - 1;$$
$$E = 3k(1 - 2\sigma_p);$$

and

$$\mu = \frac{3}{2}k\left(\frac{1 - 2\sigma_p}{1 + \sigma_p}\right).$$

The following are the common units of stress:

1 bar = 10^6 dynes per square centimeter.

= 10^5 newtons per square metre, or pascals (Pa).

= 0.1 megapascal (i.e., 0.1 × 10^6 Pa).

Thus, 10 kilobars = 1 gigapascal (*i.e.*, 10^9 Pa).

Rock Mechanics

The study of deformation resulting from the strain of rocks in response to stresses is called rock mechanics. When the scale of the deformation is extended to large geologic structures in the crust of the Earth, the field of study is known as geotectonics.

The mechanisms and character of the deformation of rocks and Earth materials can be investigated through laboratory experiments, development of theoretical models based on the properties of materials, and study of deformed rocks and structures in the field. In the laboratory, one can simulate—either directly or by appropriate scaling of experimental parameters—several conditions. Two types of pressure may be simulated: confining (hydrostatic), due to burial under rock overburden, and internal (pore), due to pressure exerted by pore fluids contained in void space in the rock. Directed applied stress, such as compression, tension, and shear, is studied, as are the effects of increased temperature introduced with depth in the Earth's crust. The effects of the duration of time and the rate of applying stress (*i.e.,* loading) as a function of time are examined. Also, the role of fluids, particularly if they are chemically active, is investigated.

Some simple apparatuses for deforming rocks are designed for biaxial stress application: a directed (uniaxial) compression is applied while a confining pressure is exerted (by pressurized fluid) around the cylindrical specimen. This simulates deformation at depth within the Earth. An independent internal pore-fluid pressure also can be exerted. The rock specimen can be jacketed with a thin, impermeable sleeve (*e.g.,* rubber or copper) to separate the external pressure medium from the internal pore fluids (if any). The specimen is typically a few centimetres in dimension.

Another apparatus for exerting high pressure on a sample was designed in 1968 by Akira Sawaoka, Naoto Kawai, and Robert Carmichael to give hydrostatic confining pressures up to 12 kilobars (1.2 gigapascal), additional directed stress, and temperatures up to a few hundred degrees Celsius. The specimen is positioned on the baseplate; the pressure is applied by driving in pistons with a hydraulic press. The end caps can be locked down to hold the pressure for time experiments and to make the device portable.

Apparatuses have been developed, typically using multianvil designs, which extend the range of static experimental conditions—at least for small specimens and limited times—to pressures as high as about 1,700 kilobars and temperatures of about 2,000 °C. Such work has been pioneered by researchers such as Peter M. Bell and Ho-Kwang Mao, who conducted studies at the Geophysical Laboratory of the Carnegie Institution in Washington, D.C. Using dynamic techniques (*i.e.,* shock from explosive impact generated by gun-type designs), even higher pressures up to 7,000 kilobars (700 gigapascal)—which is nearly twice the pressure at the centre of the Earth and seven million times greater than the atmospheric pressure at the Earth's surface—can be produced for very short times. A leading figure in such ultrapressure work is A. Sawaoka at the Tokyo Institute of Technology.

In the upper crust of the Earth, hydrostatic pressure increases at the rate of about 320 bars per kilometre, and temperature increases at a typical rate of 20°–40° C per kilometre, depending on recent crustal geologic history. Additional directed stress, as can be generated by large-scale crustal deformation (tectonism), can range up to 1 to 2 kilobars. This is approximately equal to the ultimate strength (before fracture) of solid crystalline rock at surface temperature and pressure. The stress released in a single major earthquake—a shift on a faultplane—is about 50–150 bars.

In studying the deformation of rocks one can start with the assumption of ideal behaviour: elastic strain and homogeneous and isotropic stress and strain. In reality, on a microscopic scale there are grains and pores in sediments and a fabric of crystals in igneous and metamorphic rocks. On a large scale, rock bodies exhibit physical and chemical variations and structural features. Furthermore, conditions such as extended length of time, confining pressure, and subsurface fluids affect the rates of change of deformation. Figure shows the generalized transition from brittle fracture through faulting to plastic-flow deformation in response to applied compressional stress and the progressive increase of confining pressure.

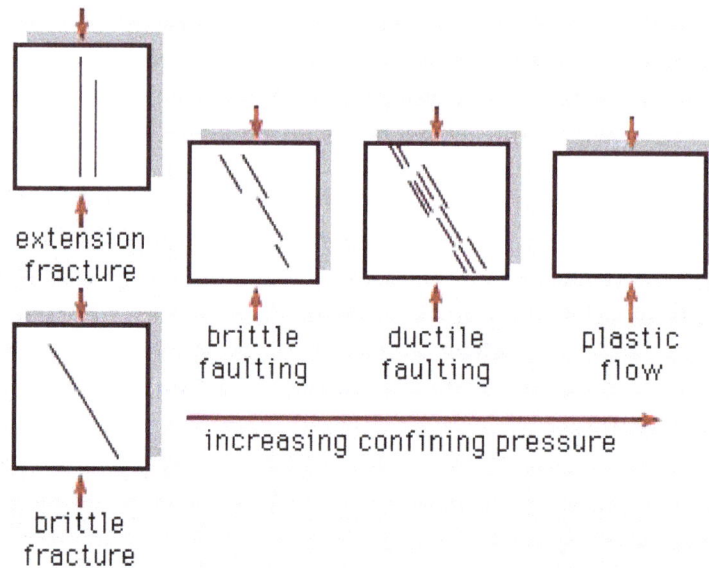

Deformation as affected by increased confining pressure.

Stress-strain Relationships

The deformation of materials is characterized by stress-strain relations. For elastic-behaviour materials, the strain is proportional to the load (*i.e.,* the applied stress). The strain is immediate with stress and is reversible (recoverable) up to the yield point stress, beyond which permanent strain results. For viscous material, there is laminar (slow, smooth, parallel) flow; one must exert a force to maintain motion because of internal frictional resistance to flow, called the viscosity. Viscosity varies with the applied stress, strain rate, and temperature. In plastic behaviour, the material strains continuously (but still has strength) after the yield point stress is reached; however, beyond this point there is some permanent deformation. In elasticoviscous deformation, there is combined elastic and viscous behaviour. The material yields continuously (viscously) for a constant applied load. An example of such behaviour is creep, a slow, permanent, and continuous deformation occurring under constant load over a long time in such materials as crystals, ice, soil and sediment, and rocks at depth. In firmoviscous behaviour, the material is essentially solid but the strain is not immediate with application of stress; rather, it is taken up and released exponentially. A plasticoviscous material exhibits elastic behaviour for initial stress (as in plastic behaviour), but after the yield point stress is reached, it flows like a viscous fluid.

The coefficient of viscosity (η) is the ratio of applied stress to the rate of straining (change of strain with time). It is measured in units of poise; one poise equals one dyne-second per square centimetre.

Rheology is the study of the flow deformation of materials. The concept of rheidity refers to the capacity of a material to flow, arbitrarily defined as the time required with a shear stressapplied for the viscous strain to be 1,000 times greater than the elastic strain. It is thus a measure of the threshold of fluidlike behaviour. Although such behaviour depends on temperature, relative comparisons can be made.

Typical stress-strain (deformation) curves for rock materials are shown in Figure. The stress σ, compression in the figure, is force per unit area. The strain ε is fractional shortening of the specimen parallel to the applied compression; it is given here in percent. The brittle material behaves elastically nearly until the point of fracture (denoted X), whereas the ductile (plastically deformable) material is elastic up to the yield point but then has a range of plastic deformation before fracturing. The ability to undergo large permanent deformation before fracture is called ductility. For plastic deformation, the flow mechanisms are intracrystalline (slip and twinning within crystal grains), intercrystalline motion by crushing and fracture (cataclasis), and recrystallization by solutioning or solid diffusion.

Typical stress-strain curves for rock materials. Each X represents the
point of fracture for the corresponding material.

If the applied stress is removed while a ductile material is in the plastic range, part of the strain is recoverable (elastically), but there is permanent deformation. The ultimate strength is the highest point (stress) on a stress-strain curve, often occurring at fracture (which is the complete loss of cohesion). The strength of a material is its resistance to failure (destruction of structure) by flow or fracture; it is a measure of the stress required to deform a body.

Effect of Environmental Conditions

The behaviour and mechanical properties of rocks depend on a number of environmental conditions. (1) Confining pressure increases the elasticity, strength (*e.g.*, yield point and ultimate fracture stress), and ductility. (2) Internal pore-fluid pressure reduces the effective stress acting on the sample, thus reducing the strength and ductility. The effective, or net, confining pressure is the external hydrostatic pressure minus the internal pore-fluid pressure. (3) Temperature lowers the strength, enhances ductility, and may enhance recrystallization. (4) Fluid solutions can enhance deformation, creep, and recrystallization. (5) Time is an influential factor as well. (6) The rate of loading (*i.e.*, the rate at which stress is applied) influences mechanical properties. (7) Compaction, as would occur with burial to depth, reduces the volume of pore space for sedimentary rocks and the crack porosity for crystalline rocks.

Rocks, which are typically brittle at the Earth's surface, can undergo ductile deformation when buried and subjected to increased confining pressure and temperature for long periods of time. If stress exceeds their strength or if they are not sufficiently ductile, they will fail by fracture—as a crystal, within a bed or rock, on an earthquake fault zone, and so on—whereas with ductility they can flow and fold.

Thermal Properties

Heat flow (or flux), q, in the Earth's crust or in rock as a building material, is the product of the temperature gradient (change in temperature per unit distance) and the material's thermal conductivity (k, the heat flow across a surface per unit area per unit time when a temperature difference exists in unit length perpendicular to the surface). Thus,

$$q = \frac{dT}{dz} \times k.$$

The units of the terms in this equation are given below, expressed first in the centimetre-gram-second (cgs) system and then in the International System of Units (SI) system, with the conversion factor from the first to the second given between them.

q, heat flow: calories per square

centime per second (cal/cm². Sec) $\times \frac{1}{23.9} \times 10^{-6}$ = watts per square metre (W/m²).

$\frac{dT}{dz}$, temperature gradient: kilometer

(°C/km, practical unit for earth) $\times 10^{-3}$ = degrees Celsius per metre (°C/m).

k, thermal conductivity: calories per square centimeter

per second per degree celsius (cal/cm² .sec. °C) $\times \frac{1}{23.9} \times 10^{-3}$ = watts per metre per degree Celsius (W/m. °C).

Thermal Conductivity

Thermal conductivity can be determined in the laboratory or in situ, as in a borehole or deep well, by turning on a heating element and measuring the rise in temperature with time. It depends on several factors: (1) chemical composition of the rock (*i.e.*, mineral content), (2) fluid content (type and degree of saturation of the pore space); the presence of water increases the thermal conductivity (*i.e.*, enhances the flow of heat), (3) pressure (a high pressure increases the thermal conductivity by closing cracks which inhibit heat flow), (4) temperature, and (5) isotropy and homogeneity of the rock.

Thermal Expansion

The change in dimension—linear or volumetric—of a rock specimen with temperature is expressed in terms of a coefficient of thermal expansion. This is given as the ratio of dimension change (*e.g.*, change in volume) to the original dimension (volume, *V*) per unit of temperature (*T*) change:

$$\frac{1}{V}\frac{\Delta V}{\Delta T}.$$

Most rocks have a volume-expansion coefficient in the range of $15–33 \times 10^{-6}$ per degree Celsius under ordinary conditions. Quartz-rich rocks have relatively high values because of the higher volume expansion coefficient of quartz.

Magnetic Properties

The magnetic properties of rocks arise from the magnetic properties of the constituent mineral grains and crystals. Typically, only a small fraction of the rock consists of magnetic minerals. It is this small portion of grains that determines the magnetic properties and magnetization of the rock as a whole, with two results: (1) the magnetic properties of a given rock may vary widely within a given rock body or structure, depending on chemical inhomogeneities, depositional or crystallization conditions, and what happens to the rock after formation; and (2) rocks that share the same lithology (type and name) need not necessarily share the same magnetic characteristics. Lithologic classifications are usually based on the abundance of dominant silicate minerals, but the magnetization is determined by the minor fraction of such magnetic mineral grains as iron oxides. The major rock-forming magnetic minerals are iron oxides and sulfides.

Although the magnetic properties of rocks sharing the same classification may vary from rock to rock, general magnetic properties do nonetheless usually depend on rock type and overall composition. The magnetic properties of a particular rock can be quite well understood provided one has specific information about the magnetic properties of crystalline materials and minerals, as well as about how those properties are affected by such factors as temperature, pressure, chemical composition, and the size of the grains. Understanding is further enhanced by information about how the properties of typical rocks are dependent on the geologic environment and how they vary with different conditions.

Applications of the Study of Rock Magnetization

An understanding of rock magnetization is important in at least three different areas: prospecting, geology, and materials science. In magnetic prospecting, one is interested in mapping the depth, size, type, and inferred composition of buried rocks. The prospecting, which may be done from ground surface, ship, or aircraft, provides an important first step in exploring buried geologic structures and may, for example, help identify favourable locations for oil, natural gas, and economic mineral deposits.

Rock magnetization has traditionally played an important role in geology. Paleomagnetic work seeks to determine the remanent magnetization and thereby ascertain the character of the Earth's field when certain rocks were formed. The results of such research have important ramifications in stratigraphic correlation, age dating, and reconstructing past movements of the Earth's crust. Indeed, magnetic surveys of the oceanic crust provided for the first time the quantitative evidence needed to cogently demonstrate that segments of the crust had undergone large-scale lateral displacements over geologic time, thereby corroborating the concepts of continental drift and seafloor spreading, both of which are fundamental to the theory of plate tectonics.

The understanding of magnetization is increasingly important in materials science as well. The design and manufacture of efficient memory cores, magnetic tapes, and permanent magnets increasingly rely on the ability to create materials having desired magnetic properties.

Basic Types of Magnetization

There are six basic types of magnetization:

1. Diamagnetism,

2. Paramagnetism,

3. Ferromagnetism,

4. Antiferromagnetism,

5. Ferrimagnetism,

6. Superparamagnetism.

Diamagnetism arises from the orbiting electrons surrounding each atomic nucleus. When an external magnetic field is applied, the orbits are shifted in such a way that the atoms set up their own magnetic field in opposition to the applied field. In other words, the induced diamagnetic field opposes the external field. Diamagnetism is present in all materials, is weak, and exists only in the presence of an applied field. The propensity of a substance for being magnetized in an external field is called its susceptibility (k) and it is defined as J/H, where J is the magnetization (intensity) per unit volume and H is the strength of the applied field. Since the induced field always opposes the applied field, the sign of diamagnetic susceptibility is negative. The susceptibility of a diamagnetic substance is on the order of -10^{-6} electromagnetic units per cubic centimetre (emu/cm³). It is sometimes denoted κ for susceptibility per unit mass of material.

Paramagnetism results from the electron spin of unpaired electrons. An electron has a magnetic dipole moment—which is to say that it behaves like a tiny bar magnet—and so when a group of electrons is placed in a magnetic field, the dipole moments tend to line up with the field. The effect augments the net magnetization in the direction of the applied field. Like diamagnetism, paramagnetism is weak and exists only in the presence of an applied field, but since the effect enhances the applied field, the sign of the paramagnetic susceptibility is always positive. The susceptibility of a paramagnetic substance is on the order of 10^{-4} to 10^{-6}emu/cm³.

Ferromagnetism also exists because of the magnetic properties of the electron. Unlike paramagnetism, however, ferromagnetism can occur even if no external field is applied. The magnetic dipole moments of the atoms spontaneously line up with one another because it is energetically favourable for them to do so. A remanent magnetization can be retained. Complete alignment of the dipole moments would take place only at a temperature of absolute zero (0 kelvin [K], or -273.15° C). Above absolute zero, thermal motions begin to disorder the magnetic moments. At a temperature called the Curie temperature, which varies from material to material, the thermally induced disorder overcomes the alignment, and the ferromagnetic properties of the substance disappear. The susceptibility of ferromagnetic materials is large and positive. It is on the order of 10 to 10^4 emu/cm³. Only a few materials—iron, cobalt, and nickel—are ferromagnetic in the strict sense of the word and have a strong residual magnetization. In general usage, particularly in engineering, the term ferromagnetic is frequently applied to any material that is appreciably magnetic.

Anti-ferromagnetism occurs when the dipole moments of the atoms in a material assume an antiparallel arrangement in the absence of an applied field. The result is that the sample has no net magnetization. The strength of the susceptibility is comparable to that of paramagnetic materials. Above a temperature called the Néel temperature, thermal motions destroy the antiparallel arrangement, and the material then becomes paramagnetic. Spin-canted (anti) ferromagnetism is a special condition which occurs when antiparallel magnetic moments are deflected from the antiferromagnetic plane, resulting in a weak net magnetism. Hematite (α-Fe_{2O_3}) is such a material.

Ferrimagnetism is an antiparallel alignment of atomic dipole moments which does yield an appreciable net magnetization resulting from unequal moments of the magnetic sublattices. Remanent magnetization is detectable. Above the Curie temperature the substance becomes paramagnetic. Magnetite (Fe_3O_4), which is the most magnetic common mineral, is a ferrimagnetic substance.

Superparamagnetism occurs in materials having grains so small (about 100 angstroms) that any cooperative alignment of dipole moments is overcome by thermal energy.

Types of Remanent Magnetization

Rocks and minerals may retain magnetization after the removal of an externally applied field, thereby becoming permanent weak magnets. This property is known as remanent magnetization and is manifested in different forms, depending on the magnetic properties of the rocks and minerals and their geologic origin and history. Delineated below are the kinds of remanent magnetization frequently observed.

CRM (chemical, or crystallization, remanent magnetization) can be induced after a crystal is formed and undergoes one of a number of physicochemical changes, such as oxidation or reduction, a phase change, dehydration, recrystallization, or precipitation of natural cements. The induction, which is particularly important in some (red) sediments and metamorphic rocks, typically takes place at constant temperature in the Earth's magnetic field.

DRM (depositional, or detrital, remanent magnetization) is formed in clastic sediments when fine particles are deposited on the floor of a body of water. Marine sediments, lake sediments, and some clay can acquire DRM. The Earth's magnetic field aligns the grains, yielding a preferred direction of magnetization.

IRM (isothermal remanent magnetization) results from the application of a magnetic field at a constant (isothermal) temperature, often room temperature.

NRM (natural remanent magnetization) is the magnetization detected in a geologic in situ condition. The NRM of a substance may, of course, be a combination of any of the other remanent magnetizations.

PRM (pressure remanent, or piezoremanent, magnetization) arises when a material undergoes mechanical deformation while in a magnetic field. The process of deformation may result from hydrostatic pressure, shock impact (as produced by a meteorite striking the Earth's surface), or directed tectonic stress. There are magnetization changes with stress in the elastic range, but the most pronounced effects occur with plastic deformation when the structure of the magnetic minerals is irreversibly changed.

TRM (thermoremanent magnetization) occurs when a substance is cooled, in the presence of a magnetic field, from above its Curie temperature to below that temperature. This form of magnetization is generally the most important, because it is stable and widespread, occurring in igneous and sedimentary rocks. TRM also can occur when dealing exclusively with temperatures below the Curie temperature. In PTRM (partial thermoremanent magnetization) a sample is cooled from a temperature below the Curie point to yet a lower temperature.

VRM (viscous remanent magnetization) results from thermal agitation. It is acquired slowly over time at low temperatures and in the Earth's magnetic field. The effect is weak and unstable but is present in most rocks.

Hysteresis and Magnetic Susceptibility

The concept of hysteresis is fundamental when describing and comparing the magnetic properties of rocks. Hysteresis is the variation of magnetization with applied field and illustrates the ability of a material to retain its magnetization, even after an applied field is removed. Figure illustrates this phenomenon in the form of a plot of magnetization (J) versus applied field (H_{ex}). J_s is the saturation (or "spontaneous") magnetization when all the magnetic moments are aligned in their configuration of maximum order. It is temperature-dependent, reaching zero at the Curie temperature. $J_{r,sat}$ is the remanent magnetization that remains when a saturating (large) applied field is removed, and J_r is the residual magnetization left by some process apart from IRM saturation, as, for example, TRM. H_c is the coercive field (or force) that is required to reduce $J_{r,sat}$ to zero, and $H_{c,r}$ is the field required to reduce J_r to zero.

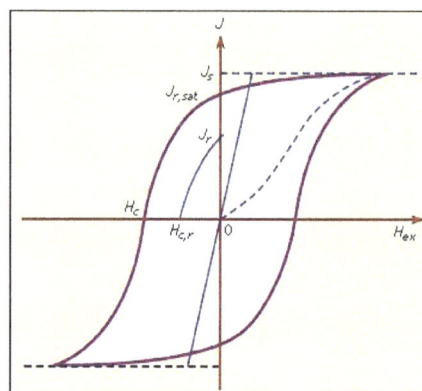

In figure, general magnetic hysteresis curve, showing magnetization (J) as a function of the external field (Hex). J_s is the saturation (or "spontaneous") magnetization; $J_{r,sat}$ is the remanent magnetization that remains after a saturating applied field is removed; J_r is the residual magnetization left by some magnetization process other than IRM saturation; H_c is the coercive field; and $H_{c,r}$ is the field necessary to reduce J_r to zero.

Magnetic susceptibility is a parameter of considerable diagnostic and interpretational use in the study of rocks. This is true whether an investigation is being conducted in the laboratory or magnetic fields over a terrain are being studied to deduce the structure and lithologic character of buried rock bodies. Susceptibility for a rock type can vary widely, depending on magnetic mineralogy, grain size and shape, and the relative magnitude of remanent magnetization present, in addition to the induced magnetization from the Earth's weak field. The latter is given as *Jinduced = kHex*, where *k* is the (true) magnetic susceptibility and *Hex* is the external (*i.e.*, the Earth's) magnetic field. If there is an additional remanent magnetization with its ratio (*Qn*) to induced magnetization being given by,

$$Q_n = \frac{J_{remanent}}{kH_{ex}},$$

Then, the total magnetization is,

$$
\begin{aligned}
J &= J_{induced} + J_{remanent} \\
&= kH_{ex} + Q_n kH_{ex} \\
&= k(1 + Q_n)H_{ex} \\
&= k_{app} H_{ex},
\end{aligned}
$$

Where, k_{app}, the "apparent" magnetic susceptibility, is $k(1 + Q_n)$.

Types of Rock

Igneous Rocks

Igneous rocks are formed from solidification and cooling of magma. This magma can be derived from partial melts of pre-existing rocks in either a planet's mantle or crust. Typically, the melting of rocks is caused by one or more of three processes namely; an increase in temperature, a decrease in pressure, or a change in composition. Igneous comes from word "ignis" meaning fire, it is therefore not surprising that igneous rocks are associated with volcanic activity and their distribution is controlled by plate tectonics. One of the appealing aspects of the plate tectonics is that it accounts for reasonably well for the variety of igneous rocks and their distribution. Divergent plates are usually associated with creation of basalts and gabbros especially in the oceanic crust e.g. in the mid-Atlantic ridges. While in the intra-continental areas you can have wide aray of rocks from basic, intermediate to the acidic rocks. In the convergent plates usually granites and andesites magmas are produced e.g. In the South America, Indonesia etc.

Igneous rocks are divided into two main categories: Plutonic (intrusive) rock and volcanic (extrusive). Plutonic or intrusive rocks result when magma cools and crystallizes slowly within the Earth's crust. A common example of this type is granite. Volcanic or extrusive rocks result from magma reaching the surface either as lava or fragmental ejecta, forming rocks such as pumice or basalt. The chemical abundance and the rate of cooling of magma typically form a sequence known as Bowen's reaction series, after the Canadian petrologist Norman L. Bowen. The Bowens reaction series explain sequences of crustal formation. The Bowens series is important because it forms basis for explaining igneous mineral and textures.

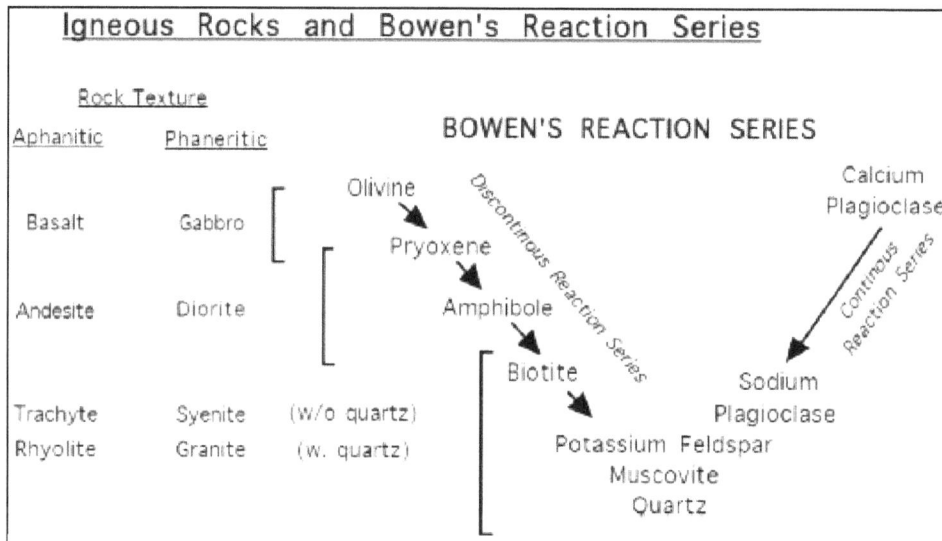

Bowens reaction series.

Types of Igneous Rocks

Igneous rocks can be differentiated according to their texture, colour and composition. The difference in these three parameters depends on environment of deposition and chemistry of magmas. Below is an explanation on of each parameter.

Texture

When magma cools slowly large crystals form and rock forms phaneritic texture on the other hand if magma cools fast then small crystals form sometime a glassy texture where no minerals form can be achieved this way. It is based on the textural difference that igneous rocks can be divided into either extrusive or intrusive rocks. Intrusive are rocks that form by magma solidifying before reaching the surface hence forming coarse grained texture while extrusive are those that magma solidify on surface forming fine grained rocks.

Colour

A rock with majorly dark minerals form mafic rocks but with more fractionation during magma cooling lighter coloured mineral are able to form based on Bowens series. Based on this colour difference the rocks can be either mafic or felsic in Figure below shows that as you move from right to left you have more ultra-mafic due to fractionation.

Composition

Igneous rocks can also be classified based on chemistry. This is mainly based on silica content as highlighted in figure below .When silica is above 75% main minerals that form are feldspars while with reduction of silica more mafic minerals form, hence basis for rock difference.

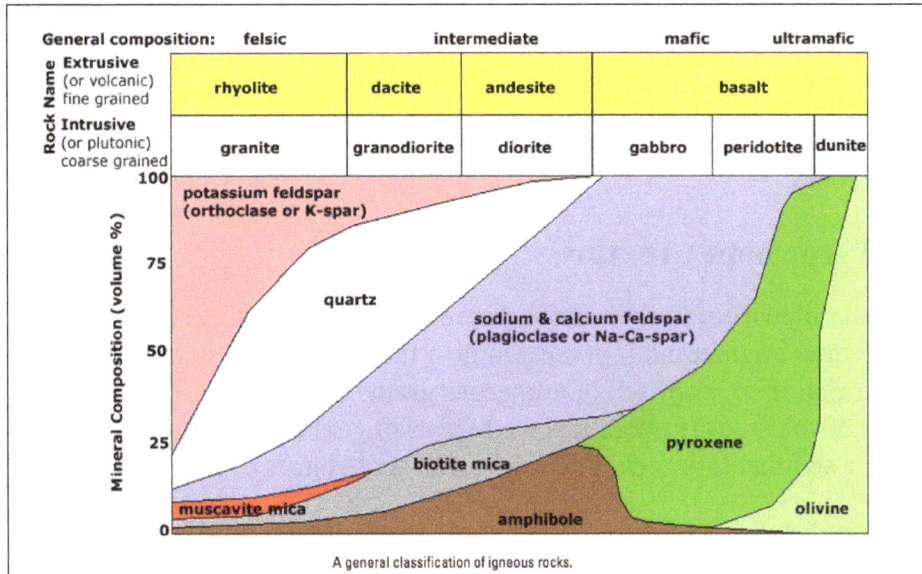

Types of igneous rocks based on texture, colour, and chemistry.

Textures of Igneous Rocks

Phaneritic Texture

Phaneritic textured rocks are comprised of large crystals that are clearly visible to the eye with or without a hand lens or binocular microscope. The entire rock is made up of large crystals, which are generally 1/2 mm to several centimetres in size; no fine matrix material is present. This texture forms by slow cooling of magma deep underground in the plutonic environment.

Aphanitic Texture

Aphanitic texture consists of small crystals that cannot be seen by the eye with or hand lens. The entire rock is made up of small crystals, which are generally less than 1/2 mm in size. This texture results from rapid cooling in volcanic or hypabyssal (shallow subsurface) environments.

Porphyritic Texture

Porphyritic rocks are composed of at least two minerals having a conspicuous (large) difference in grain size. The larger grains are termed phenocrysts and the finer grains either matrix or ground-mass. Porphyritic rocks are thought to have undergone two stages of cooling; one at depth where the larger phenocrysts formed and a second at or near the surface where the matrix grains crystallized.

Glassy Texture

Glassy textured igneous rocks are non-crystalline meaning the rock contains no mineral grains. Glass results from cooling that is so fast that minerals do not have a chance to crystallize. This may happen when magma or lava comes into quick contact with much cooler materials near the Earth's surface. Pure volcanic glass is known as obsidian.

Vesicular Texture

This term refers to vesicles (cavities) within the igneous rock. Vesicles are the result of gas expansion (bubbles), which often occurs during volcanic eruptions. Pumice and scoria are common types of vesicular rocks.

Fragmental (Pyroclastic) Texture

Pyroclastic are rocks blown out into the atmosphere during violent volcanic eruptions. These rocks are collectively termed fragmental. If you examine a fragmental volcanic rock closely you can see why. You will note that it is comprised of numerous grains or fragments that have been welded together by the heat of volcanic eruption. If you run your fingers over the rock it will often feel grainy like sandpaper or a sedimentary rock. You might also spot shards of glass embedded in the rock.

Metamorphic Rocks

Metamorphic rocks are basically rocks that have experience change due to high pressure and temperature below zone of diagenesis. Protolith refers to the original rock, prior to metamorphism In low grade metamorphic rocks, original textures are often preserved allowing one to determine the likely protolith. As the grade of metamorphism increases, original textures are replaced with metamorphic textures and other clues, such as bulk chemical composition of the rock, are used to determine the protolith. Below is an examination of the role of two agents of metamorphism.

The Role of Temperature

Changes in temperature conditions during metamorphism cause several important processes to occur. With increasing temperature, and thus higher energy, chemical bonds are able to break and reform driving the chemical reactions that changes the rock's chemistry during metamorphism. Increasing in temperature can also result in the growth of crystals. In a rock, a small number of large crystals have a higher thermodynamic stability than do a large number of small crystals. As a result, increasing temperature during metamorphism, even in the absence of any chemical change, will generally result in the amalgamation of small crystals to produce a coarser grained rock. It is a fact that individual minerals are only stable over specific temperature ranges. Thus, as temperature changes, minerals within a rock become unstable and transform through chemical reactions to new minerals. This property is very important to our interpretation of metamorphic rocks. By observing the mineral assemblage (set of minerals) within a metamorphic rock, it is often possible to make an estimate of the temperature at the time of formation. That is, minerals can be used as thermometers of the process of metamorphism.

The Role of Pressure

Pressure, the second of the two physical parameters controlling metamorphism and occurs in two forms. The most widely experienced type of pressure is lithostatic. This "rock-constant" pressure is derived from the weight of overlying rocks. Lithostatic pressure is experienced uniformly by a metamorphic rock. That is, the rock is squeezed to the same degree in all directions. Thus, there is no preferred orientation to lithostatic pressure and there is no mechanical drive to rearrange crystals within a metamorphic rock experiencing lithostatic conditions. The second pressure is the directed pressure; this is pressure of motion and action. Plate tectonics provide the underlying mechanical control for all forms of directed pressure. Thus, metamorphism is closely linked to the plate tectonic cycle and many metamorphic rocks are the products of tectonic interactions. As was the case with changes in temperature, changes in pressure, either lithostatic or directed, have important impacts upon the stability of minerals. Every mineral is stable over a range of pressures, if pressure conditions during metamorphism exceed a mineral's stability range the mineral will transform to a new phase. Many of these solid-state reactions involve polymorphic transformation – changes between minerals with the same chemistry and different crystallographic structures. Just as with temperature, mineral assemblages within a metamorphic rock can be used as a barometer to measure pressure at the time of formation.

Classification

Classification of metamorphic rocks depends on textures and its degree of metamorphism. Three kinds of criteria are normally employed in the classification of metamorphic rock. These are:

1. Mineralogical - The most abundant minerals are used as a prefix to a textural term. Thus, a schist containing biotite, garnet, quartz, and feldspar, would be called a biotite-garnet schist. A gneiss containing hornblende, pyroxene, quartz, and feldspar would be called a hornblende-pyroxene gneiss. A schist containing porphyroblasts of K-feldspar would be called a K-spar porphyroblastic schist.

2. Chemical - If the general chemical composition can be determined from the mineral assemblage, then a chemical name can be employed. For example a schist with a lot of quartz and feldspar and some garnet and muscovite would be called a garnet-muscovite quartzo-feldspathic schist. A schist consisting mostly of talc would be called a talc-magnesian schist.

3. Texture- Most metamorphic textures involve foliation. Foliation is generally caused by a preferred orientation of sheet silicates. If a rock has a slatey cleavage as its foliation, it is termed a slate, if it has a phyllitic foliation, it is termed a phyllite, if it has a shistose foliation, and it is termed a schist. A rock that shows a banded texture without a distinct foliation is termed a gneiss. All of these could be porphyroblastic (i.e. could contain porhyroblasts).A rock that shows no foliation is called a hornfels if the grain size is small, and a granulite, if the grain size is large and individual minerals can be easily distinguished with a hand lens.

Metamorphic Grade

The intensity of a metamorphic event through the use of the concept of metamorphic grade. With increasing depth in the Earth, ambient temperature and pressure conditions rise steadily. Thus, within the continental crust, temperatures vary from approximately 200 °C at 5 km to 800 °C at 35 km. While these temperatures are extreme relative to our everyday experiences, they are significantly below the melting point of most rocks. Likewise, lithostatic pressure increases with increasing depth. At 5 km the pressure is approximately 2 kilo bars, or about 2000 times atmospheric pressure. Deeper within the crust, at about 35 km, the pressure increases to some 10 kb. This trend of increasing temperature and pressure within the Earth is defined by a region of commonly encountered metamorphic conditions. Low temperature and pressure setting as low-grade metamorphism usually gneisses, while high temperature and intense pressure is known as high-grade metamorphism in schist environment.

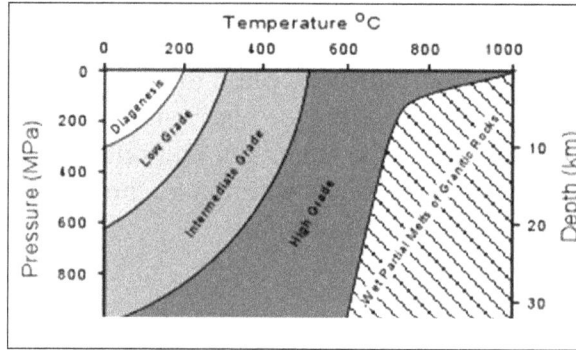

Grade of metamorphism gneiss to schist.

Types of Metamorphism

Contact Metamorphism

Contact metamorphism occurs adjacent to igneous intrusions and results from high temperatures associated with the igneous intrusion. Since only a small area surrounding the intrusion is heated by the magma, metamorphism is restricted to the zone surrounding the intrusion, called a metamorphic or contact aureole. Outside of the contact aureole, the rocks are not affected by the intrusive event. The grade of metamorphism increases in all directions toward the intrusion. Because the temperature contrast between the surrounding rock and the intruded magma is larger at shallow levels in the crust where pressure is low, contact metamorphism is often referred to as high temperature, low pressure metamorphism. The rock produced is often a fine-grained rock that shows no foliation, called a hornfels.

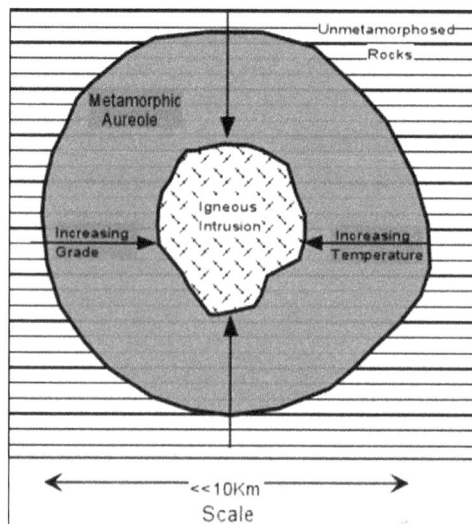

Metamorphic aureole.

Regional Metamorphism

Regional metamorphism occurs over large areas and generally does not show any relationship to igneous bodies. Most regional metamorphism is accompanied by deformation under non-hydrostatic or differential stress conditions. Thus, regional metamorphism usually results in forming metamorphic rocks that are strongly foliated, such as slates, schists, and gneisses. The differential stress usually results from tectonic forces that produce compressional stresses in the rocks, such as when two continental masses collide. Thus, regionally metamorphosed rocks occur in the cores of fold/thrust mountain belts or in eroded mountain ranges. Compressive stresses result in folding of rock and thickening of the crust, which tends to push rocks to deeper levels where they are subjected to higher temperatures and pressures.

Cataclastic Metamorphism

Cataclastic metamorphism occurs as a result of mechanical deformation, like when two bodies of rock slide past one another along a fault zone. Heat is generated by the friction of sliding along such a shear zone, and the rocks tend to be mechanically deformed, being crushed and pulverized, due to the shearing. Cataclastic metamorphism is not very common and is restricted to a narrow zone along which the shearing occurred.

Hydrothermal Metamorphism

Rocks that are altered at high temperatures and moderate pressures by hydrothermal fluids are hydrothermally metamorphosed. This is common in basaltic rocks that generally lack hydrous minerals. The hydrothermal metamorphism results in alteration to such Mg-Fe rich hydrous minerals as talc, chlorite, serpentine, actinolite, tremolite, zeolites, and clay minerals. Rich ore deposits are often formed as a result of hydrothermal metamorphism.

Burial Metamorphism

When sedimentary rocks are buried to depths of several hundred meters, temperatures greater than 300 °C may develop in the absence of differential stress. New minerals grow, but the rock does not appear to be metamorphosed. The main minerals produced are often the Zeolites. Burial metamorphism overlaps, to some extent, with diagenesis, and grades into regional metamorphism as temperature and pressure increase.

Shock Metamorphism (Impact Metamorphism)

When an extra-terrestrial body, such as a meteorite or comet impacts with the Earth or if there is a very large volcanic explosion, ultrahigh pressures can be generated in the impacted rock. These ultrahigh pressures can produce minerals that are only stable at very high pressure, such as the SiO_2 polymorphs coesite and stishovite. In addition they can produce textures known as shock lamellae in mineral grains, and such textures as shatter cones in the impacted rock.

Metamorphic Facies

The changes in mineral assemblages are due to changes in the temperature and pressure conditions of metamorphism. Thus, the mineral assemblages that are observed must be an indication of the temperature and pressure environment that the rock was subjected to. This pressure and temperature environment is referred to as Metamorphic Facies. The sequence of metamorphic facies observed in any metamorphic terrain, depends on the geothermal gradient that was present during metamorphism. Figure below highlights metamorphic faces depending on temperatures and presures. Each facies has specific index minerals as described in table.

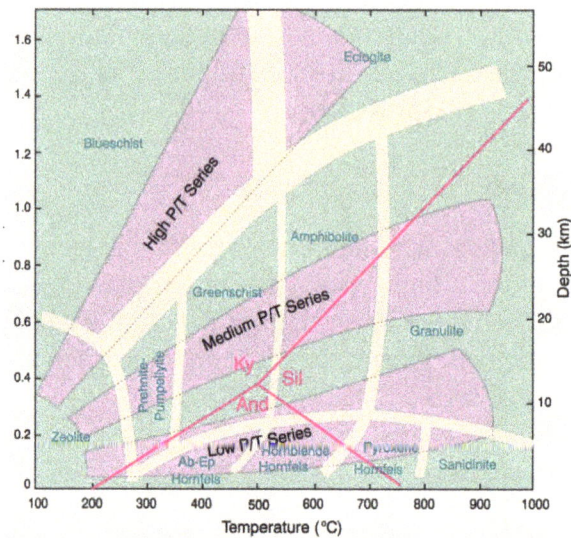

Metamorphic faces

Table: Index minerals for the metamorphic facies.

Zeolite zeolites: especially laumontite, wairakite, analcime
Prehnite-Pumpellyite prehnite + pumpellyite (+ chlorite + albite)
Greenschist chlorite + albite + epidote (or zoisite) + quartz ± actinolite
Amphibolite hornblende + plagioclase (oligoclase-andesine) ± garnet
Granulite orthopyroxene (+ clinopyrixene + plagioclase ± garnet ± hornblende)
Blueschist glaucophane + lawsonite or epidote (+albite ± chlorite)
Eclogite pyrope garnet + omphacitic pyroxene (± kyanite)
Contact facies

Igneous Activity, Metamorphism and Plate Tectonics

Plate tectonics is the mechanisms behind motion of crustal plates. It is still a controversial hypothesis though it explains very well volcanicity and igneous activity. Through this mechanism heat is transferred from deeper levels of the earth through plate margins. At these environments igneous activities and metamorphism is widespread. Figure shows the two plate tectonic environments i.e. convergent plate margins and divergent plate tectonic margins. Convergent tectonic margins are areas around ring of fire covering Indonesia, Philippines and South America in the Andes region.

Divergent plate margins are the mid-Atlantic ridge and the African Rift Valley. At these margins major geothermal resources occur e.g. in Philippines, Indonesia, Iceland and African countries within Rift valley.

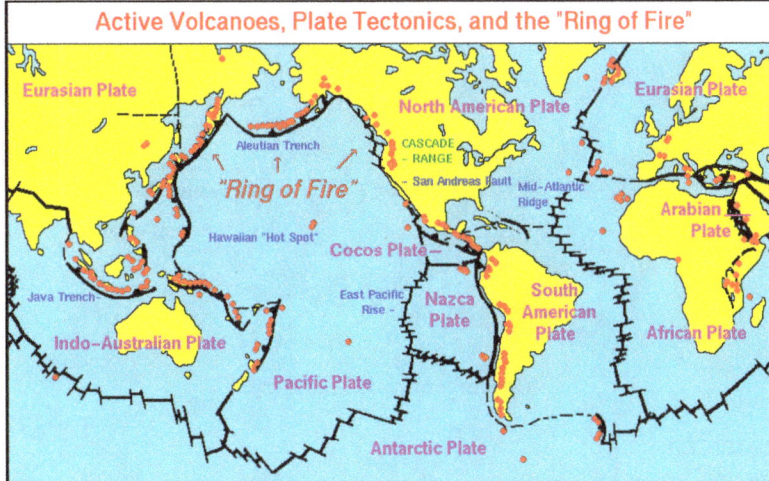

Plate tectonics and igneous provinces.

Divergent Plate Boundary

Earth plates are made of rigid lithosphere and the plastic asthenosphere. In the divergent boundaries usually basic igneous rock types are produced. Asthenosphere is usually close to the surface in the order of 5 to 10 km. It is usually plastic since temperatures are lower than those required for melting. When this zone is upwelled the pressure is reduced and melting occurs due to decompression from deeper levels hence melting. Beneath the divergent boundary the asthenosphere is welled upward and decompression occurs. At this environments Igneous of volcanic activities are pretty common and with the temperatures and pressure metamorphism also is with spread.

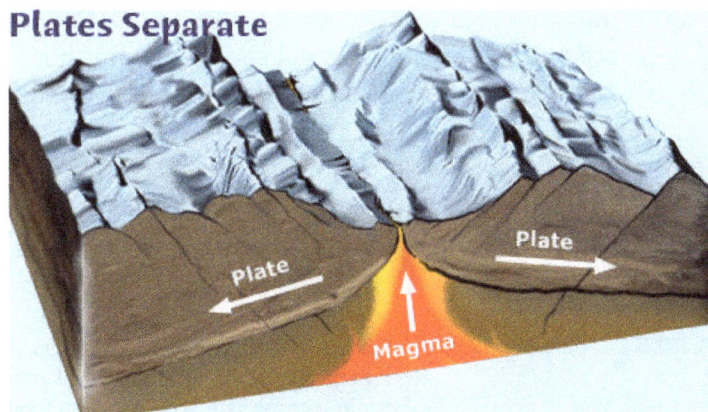

Divergent plate boundary.

Intraplate Zones

Igneous activity away from plate margin is unusual however, such hotspots exist due to mantle plumes e.g. in Hawaii, and yellow stone national park.

Convergent Plates

Intermediate and silic magmas are clearly related to convergence plate and subduction.

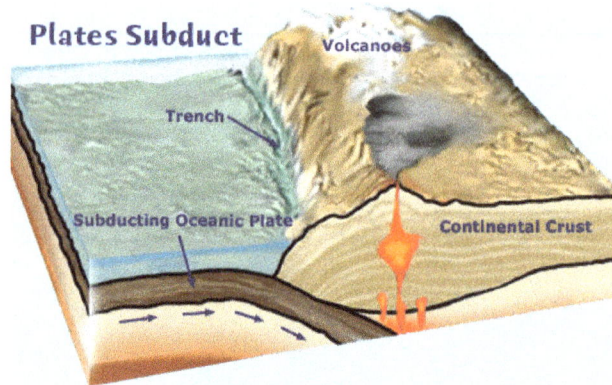

Convergent plate boundary.

Geothermal Resources

Different types of geothermal resources occur, namely. Those associated with volcanicity or igneous environments, sedimentary/geopressurised environments and to lesser extent metamorphic or dry rock system. The systems associated with igneous or volcanic environments are by far more and are high enthalpy resources making them very attractive for development. Examples of such resources are found in Kenya within igneous provinces of the Rift Valley, Iceland, Hawaii etc. Only in a few places like Hawaii are intraplate activities associated with geothermal are found.

Sedimentary Rocks

Sedimentary rocks are formed by the deposition of material at the Earth's surface and within bodies of water. Sedimentation is the collective name for processes that cause mineral and organic particles (detritus) to settle and accumulate or minerals to precipitate from a solution. Sediments can be detrital, chemical or organic sediments. Detrital sediments are mechanically eroded from pre-existing rocks. Chemical sediments on the other hand are fluid precipitates or evaporates deposited in various environments. Sedimentary rocks are important in regard to resources like limestone deposits, coal and oil. They are also important geologically in interpretation of earth's history. Below is table with descriptions of some common sedimentary rocks which include; sandstone, limestone, shale, conglomerate, and gypsum.

Rock name	Description
Limestone	The rock is classified as chemical or organic sediment depending on how it forms. It is composed primarily of calcium carbonate. The rock can form in two ways either by organically from the accumulation of shell, coral, algal and fecal debris or from chemically from the precipitation of calcium carbonate from lake or ocean water. Limestone is used in many ways. Some of the most common are: production of cement, crushed stone and acid neutralization.

Conglomerate	This is a clastic sedimentary rock that contains large (greater than 2 millimetres in diameter) rounded particles. The space between the pebbles is generally filled with smaller particles and chemical cement that bind the rock together.
Sandstone	This primarily a clastic sedimentary rock made up mainly of sand-size (1/16 to 2 millimetre diameter) weathering debris. Environments where large amounts of sand can accumulate include beaches, deserts, flood plains and deltas.
Coal	This purely organic sedimentary rock that forms mainly from plant debris. The plant debris usually accumulates in a swamp environment. Coal is combustible and is often mined for use as a fuel.

Rock Cycle

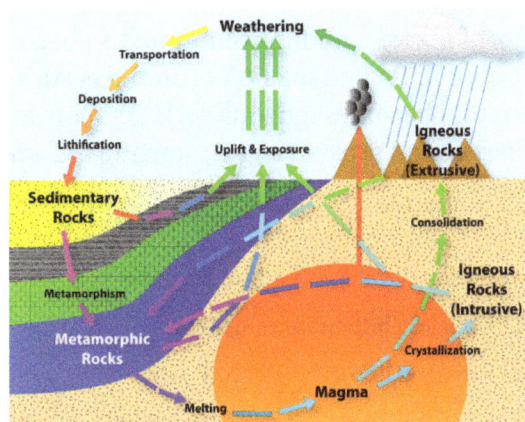

The rock cycle is an illustration that is used to explain how the three rock types are related to each other and how Earth processes change a rock from one type to another through geologic time.

Plate tectonic movement is responsible for the recycling of rock materials and is the driving force of the rock cycle.

These rocks change over hundreds of years in the six rock cycle steps:

- Weathering & Erosion: Igneous, sedimentary, and metamorphic rocks on the surface of the earth are constantly being broken down by wind and water. Wind carrying sand wears particles off rock like sandpaper. Rushing river water and crashing surf rub off all the rough edges of rocks, leaving smooth river rocks or pebbles behind. Water seeps into the cracks in mountain rocks, then freezes, causing the rocks to break open. The result of all this: large rocks are worn down to small particles. When the particles are broken off a rock and stay in the same area, it is called weathering. When the particles are carried somewhere else, it is called erosion.

- Transportation: Eroded rock particles are carried away by wind or by rain, streams, rivers, and oceans.

- Deposition: As rivers get deeper or flow into the ocean, their current slows down, and the rock particles (mixed with soil) sink and become a layer of sediment. Often the sediment builds up faster than it can be washed away, creating little islands and forcing the river to break up into many channels in a delta.

- Compaction & Cementation: As the layers of sediment stack up (above water or below), the weight and pressure compacts the bottom layers. Dissolved minerals fill in the small gaps between particles and then solidify, acting as cement. After years of compaction and cementation, the sediment turns into sedimentary rock.

- Metamorphism: Over very long periods of time, sedimentary or igneous rocks end up buried deep underground, usually because of the movement of tectonic plates. While underground, these rocks are exposed to high heat and pressure, which changes them into metamorphic rock. This tends to happen where tectonic plates come together: the pressure of the plates squish the rock that is heated from hot magma below. (Tectonic plates are large sections of the earth's crust that move separately from each other. Their movement often results in earthquakes).

- Rock Melting: Metamorphic rocks underground melt to become magma. When a volcano erupts, magma flows out of it. (When magma is on the earth's surface, it is called lava). As the lava cools it hardens and becomes igneous rock. As soon as new igneous rock is formed, the processes of weathering and erosion begin, starting the whole cycle over again.

References

- Mineralogy: newworldencyclopedia.org, Retrieved 15 June, 2019

- Mineral-chemical-compound, science: britannica.com, Retrieved 17 August, 2019

- Petrology: newworldencyclopedia.org, Retrieved 29 April, 2019

- Rock-geology, science: britannica.com, Retrieved 9 January, 2019

- Rock-cycle: mineralogy4kids.org, Retrieved 13 July, 2019

- Rock-cycle-science-lesson: homesciencetools.com, Retrieved 3 May, 2019

Chapter 5

Sedimentology

The study of sediments like sand, clay and silt along with the processes which result in their formation are studied within the discipline of sedimentology. This chapter closely examines the key aspects of these sediments to provide an extensive understanding of sedimentology.

Sedimentology encompasses the study of modern sedimentssuch as sand, mud (silt), and clay, and understanding the processes that deposit them. It also compares these observations to studies of ancient sedimentary rocks. Sedimentologists apply their understanding of modern processes to historically formed sedimentary rocks, allowing them to understand how they formed.

Sedimentary rocks cover most of the Earth's surface, record much of the Earth's history, and harbor the fossil record. Sedimentology is closely linked to stratigraphy, the study of the physical and temporal relationships between rock layers or strata. Sedimentary rocks are useful in various applications, such as for art and architecture, petroleum extraction, ceramic production, and checking reservoirs of groundwater.

Heavy minerals (dark) deposited in quartz beach sand.

Basic Principles

The aim of sedimentology, studying sediments, is to derive information on the depositional conditions that acted to deposit the rock unit, and the relation of the individual rock units in a basin into a coherent understanding of the evolution of the sedimentary sequences and basins, and thus, the Earth's geological history as a whole.

Uniformitarian geology works on the premise that sediments within ancient sedimentary rocks were deposited in the same way as sediments that are being deposited on the Earth's surface today.

In other words, the processes affecting the Earth today are the same as in the past, which then becomes the basis for determining how sedimentary features in the rock record were formed. One may compare similar features today—for example, sand dunes in the Sahara or the Great Sand Dunes National Park near Alamosa, Colorado—to ancient sandstones, such as the Wingate Sandstone of Utah and Arizona, of the southwest United States. Since both have the same features, both can be shown to have formed from aeolian (wind) deposition.

Sedimentological conditions are recorded within the sediments as they are laid down; the form of the sediments at present reflects the events of the past and all events which affect the sediments, from the source of the sedimentary material to the stresses enacted upon them after diagenesis are available for study.

The principle of superposition is critical to the interpretation of sedimentary sequences, and in older metamorphic terrains or fold and thrust beltsm where sediments are often intensely folded or deformed, recognizing younging indicators or fining up sequences is critical to interpretation of the sedimentary section and often the deformation and metamorphic structure of the region.

Folding in sediments is analyzed with the principle of original horizontality, which states that sediments are deposited at their angle of repose which, for most types of sediment, is essentially horizontal. Thus, when the younging direction is known, the rocks can be "unfolded" and interpreted according to the contained sedimentary information.

The principle of lateral continuity states that layers of sediment initially extend laterally in all directions unless obstructed by a physical object or topography.

The principle of cross-cutting relationships states that whatever cuts across or intrudes into the layers of strata is younger than the layers of strata.

Methodology

The methods employed by sedimentologists to gather data and evidence on the nature and depositional conditions of sedimentary rocks include:

- Measuring and describing the outcrop and distribution of the rock unit.

  ○ Describing the rock formation, a formal process of documenting thickness, lithology, outcrop, distribution, contact relationships to other formations.

  ○ Mapping the distribution of the rock unit, or units.

- Descriptions of rock core (drilled and extracted from wells during hydrocarbon exploration).

- Sequence stratigraphy.

  ○ Describes the progression of rock units within a basin.

- Describing the lithology of the rock.

  ○ Petrology and petrography; particularly measurement of texture, grain size, grain shape (sphericity, rounding, and so on), sorting and composition of the sediment.

- Analyzing the geochemistry of the rock.

    ◦ Isotope geochemistry, including use of radiometric dating, to determine the age of the rock, and its affinity to source regions.

Importance of Sedimentary Rocks

Sedimentary rocks provide a multitude of products that both ancient and modern societies have come to utilize.

- Art: Marble, although a metamorphosed limestone, is an example of the use of sedimentary rocks in the pursuit of aesthetics and art.

- Architectural uses: Stone derived from sedimentary rocks is used for dimension stone and in architecture, notably slate, a meta-shale, for roofing, sandstone for load-bearing buttresses.

- Ceramics and industrial materials: Clay for pottery and ceramics including bricks; cement and lime derived from limestone.

- Economic geology: Sedimentary rocks host large deposits of SEDEX ore deposits of lead-zinc-silver, large deposits of copper, deposits of gold, tungsten, and many other precious minerals, gemstones, and industrial minerals including heavy mineral sands ore deposits.

- Energy: Petroleum geology relies on the capacity of sedimentary rocks to generate deposits of petroleum oils. Coal and oil shale are found in sedimentary rocks. A large proportion of the world's uranium energy resources are hosted within sedimentary successions.

- Groundwater: Sedimentary rocks contain a large proportion of the Earth's groundwater aquifers. Human understanding of the extent of these aquifers and how much water can be withdrawn from them depends critically on knowledge of the rocks that hold them (the reservoir).

Sediment

Sediment is any particulate matter that is transported by the flow of fluids (such as water and air) and eventually deposited in a layer of solid particles. The process of deposition by settling of a suspended material is called sedimentation.

Sediments may be transported by the action of streams, rivers, glaciers, and wind. Desert sand dunes and loess (fine, silty deposits) are examples of eolian (wind) transport and deposition. Glacial moraines (rock debris) deposits and till (unsorted sediment) are ice-transported sediments. In addition, simple gravitational collapse, as occurs after the dissolution of layers of bedrock, creates sediments such as talus (slope formed by accumulated rock debris) and mountainslide deposits.

Seas, oceans, and lakes also accumulate sediment over time. The material can be terrestrial (deposited on the land) or marine (deposited in the ocean). Terrigenous deposits originate on land

and are carried by rivers and streams, but they may be deposited in terrestrial, marine, or lacustrine (lake) environments. In the mid-ocean, living organisms are primarily responsible for sediment accumulation, as their shells sink to the ocean floor after the creatures die.

Sediment builds up on human-made breakwaters because they reduce the speed of water flow, so the stream cannot carry as much sediment load.

The process of sedimentation helps renew nutrients in the soil, thereby supporting living organisms. Without such processes, the soil could become depleted of nutrients relatively quickly, and living organisms may not be able to survive in those same habitats. Moreover, deposited sediments are the source of sedimentary rocks, which can contain fossils that were covered by accumulating sediment. Lake-bed sediments that have not solidified into rock can be used to determine past climatic conditions. Thus, by analyzing sediments and sedimentary rocks, we can get glimpses of some aspects of the Earth's history.

Key Depositional Environments

Fluvial Bedforms

Rivers and streams are known as *fluvial* environments. Any particle that is larger in diameter than approximately 0.7 millimeters will form visible topographic features on the riverbed or streambed. These features, known as *bedforms*, include ripples, dunes, plane beds, and antidunes. The bedforms are often preserved in sedimentary rocks and can be used to estimate the direction and magnitude of the depositing flow.

The major fluvial environments for deposition of sediments include the following:

1. Deltas: River deltas, which are arguably intermediate between fluvial and marine environments, are landforms created by the buildup of sediment at the "mouths" of rivers and streams, that is, at places where they reach the sea. Deltas are roughly triangular in shape, but the shape depends on how the water flows, how the current changes and the amount of sediment are being carried.

2. Point bars: They are the result of an accumulation of gravel, sand, silt, and clay on the inside

bank of a bend of a river. They demonstrate a characteristic semi-ellipse shape because of the way they are formed, with larger sediment forming the base, and finer particles making up the upper part of the point bar. Point bars contribute to size and shape changes of a meander (bend) over time.

3. Alluvial fans: These are fan-shaped deposits formed where a fast-flowing stream flattens, slows, and spreads, typically at the end of a canyon onto a flatter plain.

4. Braided rivers: They consist of a network of small channels separated by small and often temporary islands called *braid bars*. Braided streams are common wherever a drastic reduction in stream gradient causes rapid deposition of the stream's sediment load.

5. Oxbow lakes: These are curved lakes formed when a wide meander (or bend) of a nearby stream or river is cut off. A combination of deposition and rapid flow work to seal the meander, cutting it off from the original body of water it was formerly connected to.

6. Levees: These are natural or artificial embankments or dikes that border the perimeter of a river. They have a wide earthen base and taper at the top. Natural levees occur as a result of tidal waves or sharp meandering of a river. Artificial levees are built to prevent flooding of the adjoining land, but they also confine the river flow, increasing the velocity of the flow.

Marine Bedforms

Marine environments (seas and oceans) also see the formation of bedforms. The features of these bedforms are influenced by tides and currents. The following are major areas for deposition of sediments in the marine environment.

1. Littoral (coastal) sands: They include beach sands, coastal bars and spits. They are largely clastic, with little faunal content.

2. The continental shelf: It consists of silty clays, with increasing content of marine fauna.

3. The shelf margin: It has a low supply of terrigenous material, mostly faunal skeletons made of calcite.

4. The shelf slope: This consists of much more fine-grained silts and clays.

5. Beds of estuaries: The resultant deposits are called "bay mud."

One other depositional environment, called the turbidite system, is a mixture of fluvial and marine environments. It is a major source of sediment for the deep sedimentary and abyssal basins, as well as for deep oceanic trenches.

Surface Runoff

Surface runoff water can pick up soil particles and transport them in overland flow for deposition at a lower land elevation or deliver that sediment to receiving waters. In this case, the sediment is usually deemed to result from erosion. If the initial impact of rain droplets dislodges soil, the phenomenon is called "splash erosion." If the effects are diffuse for a larger area and the velocity of moving runoff is responsible for sediment pickup, the process is called "sheet erosion." If there are massive gouges in the earth from high-velocity flow for uncovered soil, then "gully erosion" may result.

Rate of Sediment Settling

When a fluid (such as water) carries particles in suspension, the process by which the particulates settle to the bottom and form sediment is called settling. The term settling velocity (or fall velocity or terminal velocity (ws)) of a particle of sediment is the rate at which the particle settles in still fluid. It depends on the size, shape, and density of the grains, as well as the viscosity and density of the fluid.

For a dilute suspension of small, spherical particles in a fluid (air or water), the settling velocity can be calculated by Stoke's Law:

$$w = \frac{2(\rho_p - \rho_f)gr^2}{9\mu}$$

Where, w is the settling velocity; ρ is density (the subscripts p and f indicate particle and fluid respectively); g is the acceleration due to gravity; r is the radius of the particle; and μ is the dynamic viscosity of the fluid.

If the flow velocity is greater than the settling velocity, sediment will be transported downstream as *suspended load*.

As there will always be a range of different particle sizes in the flow, some will have sufficiently large diameters that they settle on the riverbed or streambed but still move downstream. This is known as *bed load*, and the particles are transported via such mechanisms as rolling, sliding, and "saltation" (jumping up into the flow, being transported a short distance, then settling again). Saltation marks are often preserved in solid rocks and can be used to estimate the flow rate of the rivers that originally deposited the sediments.

Erosion

One of the main causes of riverine sediment load siltation stems from "slash and burn" treatment of tropical forests. When the ground surface is stripped of vegetation and seared of all living organisms, the upper soils are vulnerable to both wind and water erosion. In a number of parts of the world, entire sectors of a country have been rendered erosive.

For example, on the Madagascar high central plateau, comprising approximately ten percent of that country's land area, virtually the entire landscape is devoid of vegetation, with gully erosive furrows typically in excess of 50 meters deep and one kilometer wide.

Shifting cultivation is a farming system that sometimes incorporates the slash and burn method in some areas of the world. The resulting sediment load in rivers is ongoing, with most rivers a dark red brown color. The accumulation of these fine particulates in the water also lead to massive fish kills, as they cover fish eggs along the bottom floor.

Sand

Sand is a natural unconsolidated granular material. Sand is composed of sand grains which range in size from 1/16 to 2 mm (62.5 ... 2000 micrometers). Sand grains are either mineral particles, rock fragments or biogenic in origin. Finer granular material than sand is referred to as silt. Coarser material is gravel. Majority of sand is dominantly composed of silicate minerals or silicate rock fragments. By far the most common mineral in sand is quartz. Hence, the term "sand" without qualification is imagined to be composed of quartz mostly. However, sand is a natural mixture which means that it is never pure. By no means can one say that quartz and sand is the same thing. Consolidated sand is a rock type known as sandstone.

Formation of Sand

Sand forms mostly by the chemical and/or physical breakdown of rocks. This process is collectively known as weathering. Physical and chemical weathering are usually treated separately, but in reality they usually go hand in hand and it is often difficult to separate one from another because they tend to support each other.

Chemical weathering is much more important sand-producing factor overall. It operates most efficiently in humid and hot climate. Physical weathering dominates in cold and dry areas. Weathering of bedrock which produces sand usually takes place in soil. Soil covers bedrock as a thin layer, providing moisture for the disintegration process of rocks.

Weathered rapakivi granite on the coast of Karelia, Russia.

Granite is a common rock type and serves as a great example of sand forming processes. Granite is composed of feldspar (pink and white) which decomposes chemically into clay minerals. Another important constituent of granite is quartz (gray). Quartz is very resistant to chemical weathering. It does not alter to any other mineral — quartz is quartz and will remain that way. It eventually goes into solutions but very slowly. Hence, disintegrated granite yields lots of quartz grains which will be transported mostly by running water as sand grains. The sample is from Italy. Width of view 21 cm.

Here is a picture of disintegrated granite (sand sample) from Sweden. It is a mixture of angular quartz and feldspar grains. The abundance of feldspar and angularity of the grains is a strong hint that this sand sample has not been transported long from its source area and the climatic conditions cannot be humid and hot. Width of sample 20 mm.

Here is an example of mature sand sample from USA (St. Peter Sandstone from the Ordovician Period). It is composed of almost pure and well-rounded quartz grains. This is what eventually happens if we give enough time for the nature to chemically destroy most other minerals that were present in the source rocks.

St. Peter Sandstone has seen much increased demand in recent years because it is well-suited for fracking purposes. Width of view 20 mm.

But what happens to other minerals? They are either converted to new stable minerals in atmospheric conditions (mostly clay minerals) or get carried away as ions in hydrous solutions and end up in the oceans. So these are freshwater rivers that carry ions to the sea and make it salty. There is a nice amount of irony in it. Clay minerals are carried by rivers also and we usually refer to this load of clay as mud. There is a muddy temporarily dry riverbed on the picture. Mud that covers these rocks is a mixture of clay minerals, fine sand, silt, and water. Barranco de las Augustias, Caldera de Taburiente, La Palma.

Composition of Sand

Sand is a residual material of preexisting rocks. It is therefore composed of minerals that were already there in the rocks before the disintegration commenced. However, there is one important aspect — sand occurs in a harsh environment where only the strongest survive.

Quartz is one of these minerals (list of minerals in sand) but not the only one. It is so dominant in most sand samples because it is so abundant. 12% of the crust is composed of it. Only feldspars are more abundant than quartz. (Here is more information about the composition of the crust).

Relatively rare minerals like tourmaline, zircon, rutile, etc. are also very resistant to weathering, but they rarely make up more than few percents of the composition of sand. These minerals are collectively referred to as heavy minerals.

Beach sand from Sri Lanka that contains lots of heavy minerals. Most of them are reddish spinel and garnet grains. Width of view 20 mm.

Heavy minerals may sometimes occur in sand in much higher concentrations. This is usually a result of hydrodynamic sorting. Either sea waves or river flow sort out heavier grains and carry lighter ones away. Such occurrences are known as placers and they are often used as a valuable mineral resource. Minerals that are often extracted from placer deposits are gold, cassiterite, ilmenite, monazite, magnetite, zircon, rutile, etc.

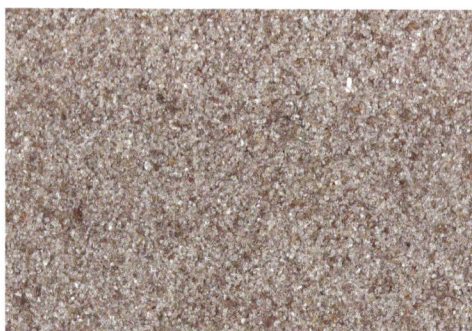

Concentrate of zircon extracted from beach sand in
South Africa. Width of view 12 mm.

Quartz definitely dominates in most sandy environments, but it is usually accompanied by feldspars. Feldspars are only moderately stable in atmospheric conditions, but their overall volume in common rocks is huge. More than half of the whole crust is composed of feldspars. Other common rock-forming minerals like amphiboles and micas also frequently occur in sand. Some common minerals in certain rocks like olivine and pyroxenes occur in sand in smaller volume because their resistance to weathering is nothing to brag about.

However, there are enough sandy beaches that are mostly composed of pyroxenes and olivine with magnetite. Such beaches with black sand occur in volcanically active areas where quartz-bearing rocks are missing.

Pyroxenes and olivine are common minerals in mafic rocks like basalt. Black sand is a typical phenomenon of oceanic volcanic islands where granite is missing and felsic quartz-rich rocks rare.

Basalt pebbles near the southern tip of La Palma slowly transforming
into black sand typical to volcanic oceanic islands.

Black sand forms in volcanic islands if quartz and biogenic grains are not available.
Here is a basaltic cliff and black sand on La Palma, Canary Islands.

Siesta Key beach sand in Florida, on the other hand, is composed almost exclusively
of quartz grains and is therefore as white as it possibly can be.

Most sand samples consist of sand grains which are composed of a single mineral — quartz grains, feldspar grains, etc. But sand may also contain grains that are an aggregate of crystals i.e. fragments of rocks (known also as lithic fragments). Lithic sand is usually immature and it also tends to form when rocks are very fine-grained. Granite usually disintegrates into distinct mineral grains, but phyllite and basalt for example are often so fine-grained that they tend to occur in sand as lithic fragments. Lithic fragments are also common in regions where erosion is rapid (mountainous terrain.

Fragments of micaceous fine-grained metamorphic rocks (phyllite, mica schist)
from Canada. Width of view 20 mm.

Sometimes sand contains new minerals or mineral aggregates that were non-existent in the source rocks. Notable example is a clay mineral glauconite which forms in marine sand and gives distinctive dark green color to many sand samples. In some instances glauconite in sand may come from disintegrated glauconitic sandstone nearby, but eventually it is of marine origin anyway.

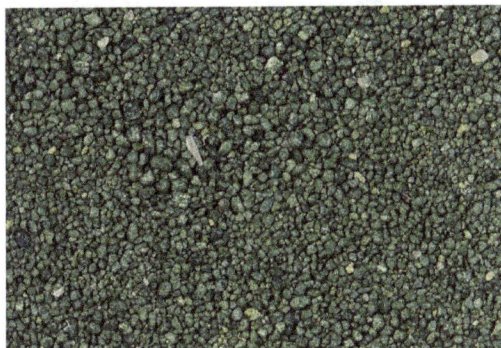

Glauconitic sand from France. Width of view 20 mm.

There are many other strange sand samples that require special formation conditions. One good example is sand in New Mexico that is composed of pure gypsum. I have written about it here: Gypsum sand. Sand with such a composition is odd and unexpected because gypsum is an evaporite mineral. It was precipitated out of hyper-saline water and it goes easily into solution again. Hence, it can only survive in dry conditions with no outlet to the sea. Halite, which is even more soluble than gypsum, is also known to form sand in special conditions.

Volcanic ash is usually treated separately, not as a type of sand. Probably because we humans tend to create artificial barriers and classification principles. We think that sand is a collection of sedimentary particles, but sedimentary and igneous rocks are two different worlds. In reality, this is more complicated because there is every reason to say that volcanic ash grains (and other pyroclastic particles like lapilli and volcanic bombs) are also sedimentary particles because they got deposited on the ground not much differently than sand grain in a dune does. Volcanic ash and sand have even comparable classification principles — volcanic ash is a pyroclastic sediment with an average grain size less than 2 millimeters. Hence, volcanic ash is a volcanic analogue of sand and silt.

Volcanic ash from St. Helens is composed of pumice fragments
and mineral grains. Width of view 20 mm.

Third major and versatile component of sand (two others were mineral grains and lithic fragments) are grains of biogenic origin. Biogenic sand is composed of fragments of exoskeletons of marine organisms. Common contributors are corals, foraminifera, sea urchins, sponges, mollusks, algae, etc. Such sand is usually known as coral sand although in many cases it contains no coral fragments at all. Biogenic sand is light-colored and widespread in low latitude marine beaches although there are exceptions. Corals indeed live only in warm water, but many other taxons can do well in colder climate (coralline algae, clams, some forams). Most biogenic sand grains are calcareous and provide material for limestone formation. Most limestones are former calcareous muds deposited on the seafloor.

Sometimes sand contains or is entirely composed of well-rounded carbonate grains that are not fragments of dead marine organisms. These grains are ooids that also require special formation conditions.

Biogenic sand from Tuamotu is mostly composed of forams. Width of view 20 mm.

Ooid sand from Cancún, Yucatán, Mexico. Width of view 5 mm.

Sand does not need to be a pure collection of either mineral, lithic, or biogenic grains. In many cases two of them and sometimes even three are mixed.

Mixture of mafic volcanic rocks and various biogenic grains in a sand from the Azores archipelago. Width of view 20 mm.

Mixture of dark-colored volcanic rocks, worn-out biogenic grains, and some silicate grains from Jeju-do Island, South Korea. Width of view 20 mm.

Texture and Transport of Sand

Geologists describe sand by measuring the roundness of grains and the distribution of grain sizes. By doing that they hope to shed some light on the origin of the grains being measured. Roundedness usually gives information about the length of transport route and distribution of grain sizes helps to determine from which environment these grains come from. River sand is usually poorly sorted and compositionally immature. Beach sand is more rounded and eolian dune sand is generally well sorted.

Poorly sorted river sand from Sikkim, India. Width of view 20 mm.

Eolian sand from the Erg Murzuk, Libya. Dune sand is generally well sorted
(grains are similar in size). Width of view 15 mm.

The average size of grains is determined by the energy of the transport medium. Higher current velocity (either stream flow or sea waves) can carry heavier load. Coarse-grained sediments therefore reveal that they were influenced by energetic medium because finer material is carried away.

Sometimes river flow is so energetic that sand grains are all carried away and only large rounded stones remain. Such lithified deposits of former riverbeds are known as conglomerates.

Sand is mostly transported by rivers, but average sand grains are too large and heavy for Average River to carry them in suspension. Hence, sand grains tend to move in jumps. They are lifted up by more energetic current and settle out when current velocity decreases and then wait for the next jump.

This mode of movement is known as saltation. Average silt grain moves differently. It is lightweight enough to be carried in suspension for a long time and this is actually one of the most important reasons why we treat silt separately from sand.

Most sand grains carried by the rivers are eventually deposited at the rivermouths where the current velocity suddenly drops. Then sea waves (longshore currents) take over and carry sand along the coastline. Sand grains carried by the rivers are also deposited on alluvial flood plains and point bars (inside bend of streams where current flow is the slowest).

Sand is also transported by wind, ocean currents, glaciers, turbidity currents, etc. Moving sand forms landforms like ripples and dunes.

Clay

Clay is a term used to describe a group of fine-grained, silicate minerals known as aluminum phyllosilicates, containing variable amounts of chemically associated water. Clay is plastic when wet, which means it can be easily shaped. When dry, it becomes firm, and when subject to high temperature, known as firing, permanent physical and chemical changes occur. These changes cause the clay to be hardened. A fireplace or oven specifically designed for hardening clay is called a kiln.

Clay soils are distinguished from other types of soil such as silt by the small grain size, flake or layered shape, affinity for water, and high degree of plasticity. The grain size of clay is typically less than two micrometers (μm) in diameter. Depending on the content of the soil, clay can appear in various colors, from a dull gray to a deep orange-red.

These cliffs in Martha's Vineyard, Massachusetts, are made almost entirely of clay.

People discovered the useful properties of clay in prehistoric times, and one of the earliest artifacts ever uncovered is a drinking vessel made of sun-dried clay. Clays remain among the cheapest and most widely used materials, to make items ranging from art objects to bricks and cookware. They are also used in industrial processes such as papermaking and cement production. An open mine for extracting clay is called a *clay pit*.

Clay Minerals

Clay minerals are rich in silicon and aluminum oxides and hydroxides, and sometimes contain variable amounts of iron, magnesium, alkali metals, alkaline earths, and other cations. Clays have structures similar to the micas and therefore form flat, hexagonal sheets.

Clays are generally formed by the chemical weathering of silicate-bearing rocks by carbonic acid, but some are formed by hydrothermal activity. Clay minerals are common in fine-grained sedimentary rocks such as shale, mudstone, and siltstone, and in fine-grained metamorphic slate and phyllite.

Clay minerals include the following groups:

- Kaolinite group, which includes the minerals kaolinite, dickite, halloysite, and nacrite.

  ◦ Some sources include the serpentine group, based on structural similarities.

- Smectite group, which includes pyrophyllite, talc, vermiculite, sauconite, saponite, nontronite, and montmorillonite.

- Illite group, which includes the clay-micas. Illite is the only common mineral in this group.

- Chlorite group, which includes a wide variety of similar minerals with considerable chemical variation. This group is not always considered a part of the clays and is sometimes classified as a separate group within the phyllosilicates.

There are about 30 different types of 'pure' clays in these categories, but most clay in nature is mixtures of these different types, along with other weathered minerals.

Varve (or *varved clay*) is clay with visible annual layers, formed by seasonal differences in erosion and organic content. This type of deposit is common in former glacial lakes from the Ice Age.

Quick clay is a unique type of marine clay, indigenous to the glaciated terrains of Norway, Canada, and Sweden. It is highly sensitive clay, prone to liquefaction, and it has been involved in several deadly landslides.

Structure

Like all phyllosilicates, clay minerals are characterised by two-dimensional *sheets* of corner-sharing tetrahedra made of SiO_4 and AlO_4. Each tetrahedron shares three of its vertex oxygen atoms with other tetrahedra. The fourth vertex is not shared with another tetrahedron and all of the tetrahedra "point" in the same direction—in other words, all the unshared vertices lie on the same side of the sheet. These tetrahedral sheets have the chemical composition $(Al,Si)_3O_4$.

In clays, the tetrahedral sheets are always bonded to octahedral sheets. The latter are formed from small cations, such as aluminum or magnesium cations, coordinated by six oxygen [atom|atoms]]. The unshared vertex from the tetrahedral sheet also forms part of one side of the octahedral sheet, but an additional oxygen atom is located above the gap in the tetrahedral sheet at the center of the six tetrahedra. This oxygen atom is bonded to a hydrogen atom forming an OH (hydroxide) group in the clay structure.

Clays can be categorized according to the way that the tetrahedral and octahedral sheets are packaged into "layers." If each layer consists of only one tetrahedral and one octahedral group, the clay is known as a 1:1 clay. Likewise, a 2:1 clay has two tetrahedral sheets, with the unshared vertex of each sheet pointing toward each other and forming each side of the octahedral sheet.

Depending on the composition of the tetrahedral and octahedral sheets, the layer will have no electriccharge or will have a net negative charge. If the layers are charged, this charge is balanced by interlayer cations such as Na^+ or K^+. In each case the interlayer can also contain water. The crystal structure is formed from a stack of layers interspaced with the interlayers.

Uses of Clay

The properties of clay make it an ideal material for producing durable pottery items for both practical and decorative purposes. By using different types of clay and firing conditions, one can produce earthenware, stoneware, and porcelain.

Quaternary clay in Estonia.

Clays sintered in fire were the first type of ceramic. They continue to be widely used, to produce such items as bricks, cooking pots, art objects, and dishware. Even some musical instruments, such as the ocarina, are made with clay. Industrial processes that involve the use of clay include papermaking, cement production, pottery manufacture, and chemical filtration.

Silt

Silt is solid, dust-like sediment that water, ice, and wind transport and deposit. Silt is made up of rock and mineral particles that are larger than clay but smaller than sand. Individual silt particles are so small that they are difficult to see. To be classified as silt, a particle must be less than .005 centimeters (.002 inches) across. Silt is found in soil, along with other types of sediment such as clay, sand, and gravel.

Silty soil is slippery when wet, not grainy or rocky. The soil itself can be called silt if its silt content is greater than 80 percent. When deposits of silt are compressed and the grains are pressed together, rocks such as siltstone form.

Silt is created when rock is eroded, or worn away, by water and ice. As flowing water transports tiny rock fragments, they scrape against the sides and bottoms of stream beds, chipping away more rock. The particles grind against each other, becoming smaller and smaller until they are silt-size. Glaciers can also erode rock particles to create silt. Finally, wind can transport rock particles through a canyon or across a landscape, forcing the particles to grind against the canyon wall or one another. All three processes create silt.

Silt can change landscapes. For example, silt settles in still water. So, deposits of silt slowly fill in places like wetlands, lakes, and harbors. Floods deposit silt along river banks and on flood plains. Deltas develop where rivers deposit silt as they empty into another body of water. About 60 percent of the Mississippi River Delta is made up of silt.

In some parts of the world, windblown silt blankets the land. Such deposits of silt are known as loess. Loess landscapes, such as the Great Plains, are usually a sign of past glacial activity.

Many species of organisms thrive in slick, silty soil. Lotus plants take root in muddy, silty wetlands, but their large, showy flowers blossom above water.

Many species of frog hibernate during the cold winter by burying themselves in a layer of soft silt at the bottom of a lake or pond. Water at the bottom of a body of water does not freeze, and the silt provides some insulation, or warmth, for the animal.

Silty soil is usually more fertile than other types of soil, meaning it is good for growing crops. Silt promotes water retention and air circulation. Too much clay can make soil too stiff for plants to thrive. In many parts of the world, agriculture has thrived in river deltas, where silt deposits arc rich, and along the sides of rivers where annual floods replenish silt. The Nile River Delta in Egypt is one example of an extremely fertile area where farmers have been harvesting crops for thousands of years.

When there aren't enough trees, rocks, or other materials to prevent erosion, silt can accumulate quickly. Too much silt can upset some ecosystems.

"Slash and burn" agriculture, for instance, upsets the ecosystem by removing trees. Agricultural soils are washed away into rivers, and nearby waterways are clogged with silt. Animals and plants that have adapted to live in moderately silty soil are forced to find a new niche in order to survive. The river habitats of some organisms in the Amazon River, such as the pink Amazon River dolphin, also called the boto, are threatened. River dolphins cannot locate prey as well in silty water.

Agricultural and industrial runoff can also clog ecosystems with silt and other sediment. In areas that use chemical fertilizers, runoff can make silt toxic. Toxic silt can poison rivers, lakes, and streams. Silt can also be made toxic by exposure to industrial chemicals from ships, making the silt at the bottom of ports and harbors especially at risk. When the city of Melbourne, Australia, decided to deepen its harbor in 2008, many people worried that disturbing millions of tons of silt, filled with chemicals like arsenic and lead, would threaten the waterway's ecosystem.

References

- Sedimentology: newworldencyclopedia.org, Retrieved 18 April, 2019

- Sediment: newworldencyclopedia.org, Retrieved 28 July, 2019

- Sand: sandatlas.org, Retrieved 20 February, 2019

- Clay: newworldencyclopedia.org, Retrieved 11 May, 2019

- Silt: nationalgeographic.org, Retrieved 10 March, 2019

Chapter 6

Hydrogeology

The domain of geology which studies the movement and distribution of groundwater in the soil and rocks of the Earth's crust is called hydrogeology. The key areas of study within this discipline include aquifers and the hydrological cycle. This chapter has been carefully written to provide an easy understanding of these facets of hydrogeology.

Hydrogeology is the study of water contained in materials of Earth's crust, the physical and chemical characteristics of this water, its origin, evolution, and ultimate destination. Hydrogeology is the term used by geologists and hydrogeologists for this study. Geohydrology is the term most often used by engineers. The two terms are roughly equivalent.

The water contained in materials of Earth's crust is called groundwater (sometimes spelled as "ground water" to distinguish ground water from surface water). Groundwater is sometimes defined as water below the earth's surface, but groundwater may occur at the surface especially after heavy rainfall in certain areas.

When groundwater is not at the earth's surface, there is a zone beneath the surface where the majority of pore (open) spaces are filled with air. This is called the vadose zone or zone of aeration. Below a certain depth, all the pore spaces are filled with water. This is known as the phreatic zone or zone of saturation. The zone of saturation extends downward until pressure of the overlying materials is so great that there are no pore spaces available. The area separating the vadose and phreatic zones is called the water table, which is usually represented on cross sections as a dashed line. The dashed line, as opposed to a solid line, indicates that the water table moves up and down with the seasons, being higher and nearer Earth's surface during wet seasons and lower and deeper below Earth's surface during dry seasons.

Modern groundwater studies have their origin in the middle 1850s when a French engineer, Henry Darcy, published a report describing an experiment he conducted with a tube on an incline that he had filled with sand. Darcy's experiments led to the first quantitative "law" in hydrogeology, used to determine the rate of flow of groundwater and now known as Darcy's law. It can be expressed mathematically as v = KIA, where "v" equals the rate of groundwater flow, "K" equals hydraulic conductivity or permeability, "A" equals the area of a cross-section of the water-bearing unit (e.g., cross-section of Darcy's cylinder of sand) and "I" equals the hydraulic gradient.

Hydraulic conductivity or permeability (K) is measured from the material through which the groundwater is flowing and has the dimensions of length per unit time (L/T, e.g., cm per second, feet per year, etc.).

Groundwater occurs underground in bodies of Earth materials called aquifers, which may be of two types: unconfined or confined. Unconfined aquifers are bound at their top by the water table. Confined aquifers are bound both top and bottom by materials through which little or no water flows,(i.e., impermeable materials), or aquicludes (also known as aquitards). The materials holding the water are usually inclined to the horizontal so that pressure builds up in the aquifer. When the aquifer is drilled, the water rises to the highest level of water confined within the aquifer or sometimes to the surface.

Groundwater

Groundwater or ground water is water located within the ground's zone of saturation, where the soil pore spaces and fractures in the rock are completely filled with water. It differs from soil water, which is the water that is in the unsaturated zone, or zone of aeration, where the soil pore spaces contain air and water but are not completely saturated. The term groundwater also has been used more broadly as any water beneath the earth's surface in soil.

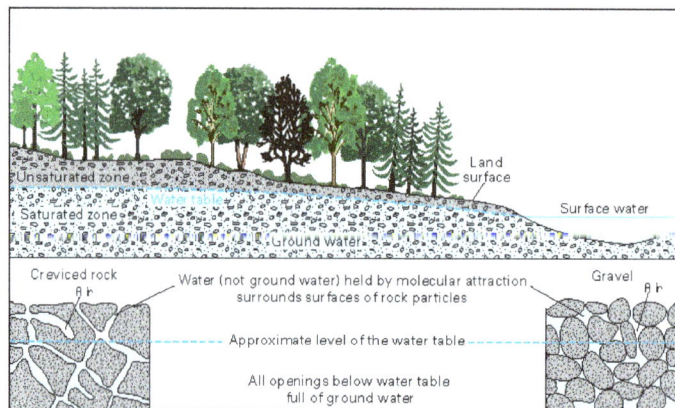

Groundwater is located within the zone of saturation

The depth at which soil pore spaces or fractures and voids in rock become completely saturated with water is called the water table; in other words, below the level of the water table the soil pores and rock fractures are saturated with water. An aquifer is a layer within the zone of saturation that can readily yield and store water, such as in interconnected spaces (fractures, cracks, poor spaces, etc.) that can provide a source of water for a well.

As part of the hydrologic cycle, groundwater stores and transmits water that has filtered down from the surface and it also slowly flows back to the surface, with natural discharge at places such as springs, seeps, and wetlands. Groundwater discharging into a stream provides water to allow the stream to flow throughout the year. Groundwater also is withdrawn for agricultural, municipal, and industrial use by constructing and operating extraction wells.

Although a vitally important renewable resource, that serves many critical economic and environmental needs, groundwater reserves in various regions face such threats as depletion from overdraft and contamination.

Importance of Groundwater

Groundwater is a renewable resource that serves many critical economic and environmental needs.

Economically, it is the source of drinking water for many communities (about half the population in the United States and nearly all the rural population), as well as providing water for agricultural and industrial needs.

Groundwater is also ecologically important. The importance of groundwater to ecosystems is often overlooked. Groundwaters sustain streams, wetlands, and lakes, as well as subterranean ecosystems within karst or alluvial aquifers. While a rain storm or snow melt can provide a lot of water for a stream, at other times of the year the stream is provided all the water by groundwater seeping through stream banks and stream beds (called base flow), allowing the streams to flow year round (Stevens).

Not all ecosystems need groundwater, of course. Some terrestrial ecosystems—for example, those of the open deserts and similar arid environments—exist on irregular rainfall and the moisture it delivers to the soil, supplemented by moisture in the air. While there are other terrestrial ecosystems in more hospitable environments where groundwater plays no central role, groundwater is in fact fundamental to many of the world's major ecosystems. Water flows between groundwaters and surface waters. Most rivers, lakes, and wetlands are fed by, and (at other places or times) feed groundwater, to varying degrees. Groundwater feeds soil moisture through percolation, and many terrestrial vegetation communities depend directly on either groundwater or the percolated soil moisture above the aquifer for at least part of each year. Hyporheic zones (the mixing zone of streamwater and groundwater) and riparian zones are examples of ecotones largely or totally dependent on groundwater.

Issues

Two key issues facing groundwater reserves are:

1. Depletion of groundwater

2. Contamination

Groundwater is depleted as is pumped out and used faster than it is replenished. This can have the effect of lowering the water table, which in turn can cause drying up of wells and the need for a well owner to deepen the well, lower the pump, or drill a new well, and greater energy costs for operation a pump; reduction of water that goes back into streams and lakes and loss of wildlife habitat and vegetation; and land subsidence. This last issue can arise when the loss of water causes soil to compact, collapse, and drop, and thus the loss of support below ground for structures on the surface.

Groundwater contamination can occur from a number of sources. Toxins can filter down and waste from landfills and agricultural runoff. As water tables are lowered, saltwater contamination can increase, as the freshwater/saltwater boundary is disrupted and saltwater migrates inward as well as upward from the saline groundwater.

Furthermore, as water moves through the landscape, it collects soluble salts, mainly sodium chloride. As the water enters the atmosphere through evapotranspiration, these salts are left behind. In irrigation districts, poor drainage of soils and surface aquifers can result in water tables' coming to the surface in low-lying areas. Major land degradation problems of soil salinity and waterlogging result, combined with increasing levels of salt in surface waters. As a consequence, major damage has occurred to local economies and environments.

Unlike river waters being overused and polluted, groundwater problems are less evident, as aquifers are out of sight. Another problem is that water management agencies, when calculating the "sustainable yield" of aquifer and river water, have often counted the same water twice, once in the aquifer, and once in its connected river. This problem, although understood for centuries, has persisted, partly through inertia within government agencies.

In general, the time lags inherent in the dynamic response of groundwater to development have been ignored by water management agencies, decades after scientific understanding of the issue was consolidated. In brief, the effects of groundwater overdraft (although undeniably real) may take decades or centuries to manifest themselves. In a classic study in 1982, Bredehoeft and colleagues modeled a situation where groundwater extraction in an intermontane basin withdrew the entire annual recharge, leaving "nothing" for the natural groundwater-dependent vegetation community. Even when the borefield was situated close to the vegetation, 30% of the original vegetation demand could still be met by the lag inherent in the system after 100 years. By year 500, this had reduced to 0%, signalling complete death of the groundwater-dependent vegetation. The science has been available to make these calculations for decades; however, in general water management agencies have ignored effects that will appear outside the rough time frame of political elections. Sophocleous argues that management agencies must define and use appropriate time frames in groundwater planning. This will mean calculating groundwater withdrawal permits based on predicted effects decades, sometimes centuries in the future.

Overdraft

Wetlands contrast the arid landscape around Middle Spring,
Fish Springs National Wildlife Refuge, Utah.

Over-use of groundwater, known as *overdraft*, can lead to depletion and cause major problems to human users and to the environment. The most evident problem (as far as human groundwater use is concerned) is a lowering of the water table beyond the reach of existing wells.

Subsidence

Subsidence occurs when too much water is pumped out from underground, deflating the space below the above-surface, and thus causing the ground to collapse. The result can look like craters on plots of land. This occurs because, in its natural equilibrium state, the hydraulic pressure of groundwater in the pore spaces of the aquifer and the aquitard supports some of the weight of the overlying sediments. When groundwater is removed from aquifers by excessive pumping, pore pressures in the aquifer drop and compression of the aquifer may occur. This compression may be partially recoverable if pressures rebound, but much of it is not. When the aquifer gets compressed, it may cause land subsidence, a drop in the ground surface.

Pollution

Iron oxide staining caused by reticulation from an
unconfined aquifer in karst topography. Perth, Western Australia.

Water pollution of groundwater, from pollutants released on the surface that can work their way down into groundwater, can create a contaminant plume within an aquifer. Movement of water and dispersion within the aquifer spreads the pollutant over a wider area, its advancing boundary often called a plume edge, which can then intersect with groundwater wells or emerge into surface water via such means as seeps and springs, making the water supplies unsafe for humans and wildlife. The interaction of groundwater contamination with surface waters is analyzed by use of hydrology transport models.

The stratigraphy of the area plays an important role in the transport of these pollutants. An area can have layers of sandy soil, fractured bedrock, clay, or hardpan. Areas of karst topography on limestone bedrock are sometimes vulnerable to surface pollution from groundwater. Earthquake faults also can be entry routes for downward contaminant entry. Water table conditions are of great importance for drinking water supplies, agricultural irrigation, waste disposal (including nuclear waste), wildlife habitat, and other ecological issues.

In the United States, upon commercial real estate property transactions both groundwater and soil are the subjects of scrutiny, with a Phase I Environmental Site Assessment normally being prepared to investigate and disclose potential pollution issues. In the San Fernando Valley of California, real estate contracts for property transfer below the Santa Susana Field Laboratory (SSFL) and eastward have clauses releasing the seller from liability for groundwater contamination consequences from existing or future pollution of the Valley Aquifer.

Love Canal was one of the most widely known examples of groundwater pollution. In 1978, residents of the Love Canal neighborhood in upstate New York noticed high rates of cancer and an alarming number of birth defects. This was eventually traced to organic solvents and dioxins from an industrial landfill that the neighborhood had been built over and around, which had then infiltrated into the water supply and evaporated in basements to further contaminate the air. Eight hundred families were reimbursed for their homes and moved, after extensive legal battles and media coverage.

Another example of widespread groundwater pollution is in the Ganges Plain of northern India and Bangladesh where severe contamination of groundwater by naturally occurring arsenic affects 25% of water wells in the shallower of two regional aquifers. The pollution occurs because aquifer sediments contain organic matter that generates anaerobic conditions in the aquifer. These conditions result in the microbial dissolution of iron oxides in the sediment and, thus, the release of the arsenic, normally strongly bound to iron oxides, into the water.

Groundwater Flow

In the unsaturated zone the region between the ground surface and the water table water percolates straight down, like the water passing through a drip coffee maker, for this water moves only in response to the downward pull of gravity. But in the zone of saturation the region below the water table water flow is more complex, for in addition to the downward pull of gravity, water responds to differences in pressure. Pressure can cause groundwater to flow sideways, or even upward. Thus, to understand the nature of groundwater flow, we must first understand the origin of pressure in groundwater. For simplicity, we'll consider only the case of groundwater in an unconfined aquifer.

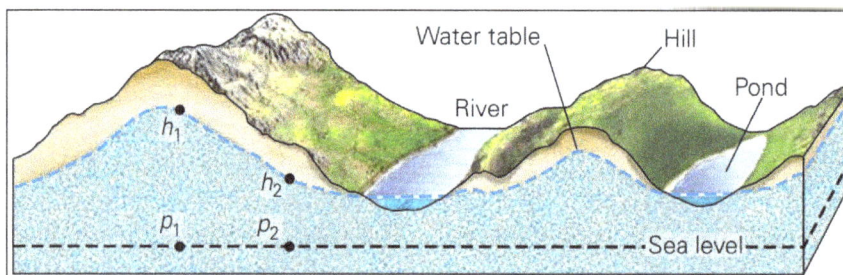

The shape of water table beneath hilly topography.

Pressure in groundwater at a specific point underground is caused by the weight of all the overlying water from that point up to the water table. (The weight of overlying rock does not contribute to the pressure exerted on groundwater, for the contact points between mineral grains bear the rock's weight). Thus, a point at a greater depth below the water table feels more pressure than does a point at lesser depth. If the water table is horizontal, the pressure acting on an imaginary horizontal reference plane at a specified depth below the water table is the same everywhere. But if the water table is not horizontal, as shown in above, the pressure at points on a horizontal reference plane at depth changes with location. For example, the pressure acting at point p1, which lies below the hill in figure above, is greater than the pressure acting at point p2, which lies below the valley, even though both p1 and p2 are at the same elevation.

Both the elevation of a volume of groundwater and the pressure within the water provide energy that, if given the chance, will cause the water to flow. Physicists refer to such stored energy as potential energy. The potential energy available to drive the flow of a given volume of groundwater at a location is called the hydraulic head. To measure the hydraulic head at a point in an aquifer, hydrogeologists drill a vertical hole down to the point and then insert a pipe in the hole. The height above a reference elevation (for example, sea level) to which water rises in the pipe represents the hydraulic head water rises higher in the pipe where the head is higher. As a rule, groundwater flows from regions where it has higher hydraulic head to regions where it has lower hydraulic head. This statement generally implies that groundwater regionally flows from locations where the water table is higher to locations where the water table is lower.

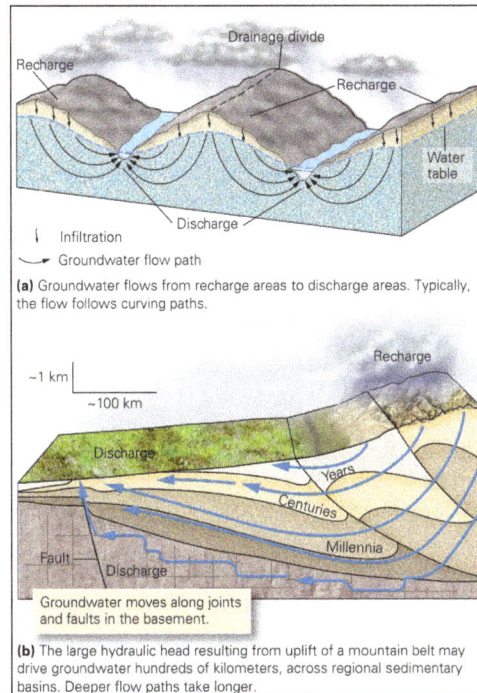

(a) Groundwater flows from recharge areas to discharge areas. Typically, the flow follows curving paths.

(b) The large hydraulic head resulting from uplift of a mountain belt may drive groundwater hundreds of kilometers, across regional sedimentary basins. Deeper flow paths take longer.

Hydrogeologists have calculated how hydraulic head changes with location underground, by taking into account both the effect of gravity and the effect of pressure. These calculations reveal that groundwater flows along concave-up curved paths, as illustrated in cross section. These curved paths eventually take groundwater from regions where the water table is high (under a hill) to regions where the water table is low (below a valley), but because of flow-path shape, some groundwater may flow deep down into the crust along the first part of its path and then may flow back up, toward the ground surface, along the final part of its path. The location where water enters the ground (where the flow direction has a downward trajectory) is called the recharge area, and the location where groundwater flows back up to the surface is called the discharge area.

Flowing water in an ocean current moves at up to 3 km per hour, and water in a steep river channel can reach speeds of up to 30 km per hour. In contrast, groundwater moves at less than a snail's pace, between 0.01 and 1.4 m per day (about 4 to 500 m per year). Groundwater moves much more slowly than surface water, for two reasons. First, groundwater moves by percolating through a complex, crooked network of tiny conduits, so it must travel a much greater distance than it would if it could follow a straight path. Second, friction between groundwater and conduit walls slows down the water flow.

Simplistically, the velocity of groundwater flow depends on the slope of the water table and the permeability of the material through which the groundwater is flowing. Thus, groundwater flows faster through high-permeability rocks than it does through low-permeability rocks and it flows faster in regions where the water table has a steep slope than it does in regions where the water table has a gentle slope. For example, groundwater flows relatively slowly (2 m per year) through a low-permeability aquifer under the Great Plains, but flows relatively quickly (30 m per year) through a high-permeability aquifer under a steep hillslope. In detail, hydrogeologists use Darcy's Law to determine flow rates at a location.

Darcy's Law for Groundwater Flow

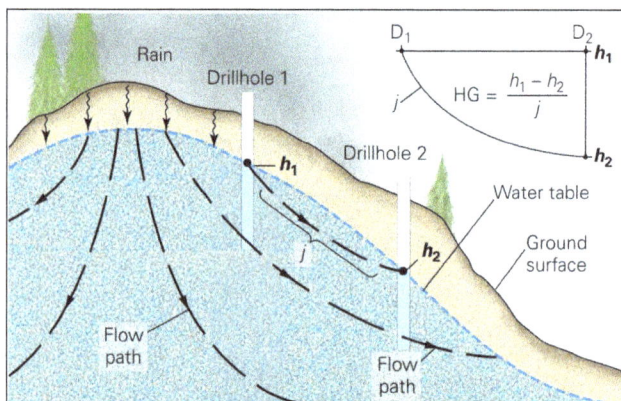

The level to which water rises in a drill hole is the hydraulic head (h). The hydraulic gradient (HG) is the difference in head divided by the length of the flow path.

The rate at which groundwater flows at a given location depends on the permeability of the material containing the groundwater; groundwater flows faster in a more permeable material than it does in a less permeable material. The rate also depends on the hydraulic gradient, the change in hydraulic head per unit of distance between two locations, as measured along the flow path.

To calculate the hydraulic gradient, we divide the difference in hydraulic head between two points by the distance between the two points as measured along the flow path. This can be written as a formula:

Hydraulic gradient = $h_1 - h_2/j$

where, $h_1 - h_2$ is the difference in head (given in meters or feet, because head can be represented as an elevation) between two points along the water table, and j is the distance between the two points as measured along the flow path. A hydraulic gradient exists anywhere that the water table has a slope. Typically, the slope of the water table is so small that the path length is almost the same as the horizontal distance between two points. So, in general, the hydraulic gradient is roughly equivalent to the slope of the water table.

In 1856, a French engineer named Henry Darcy carried out a series of experiments designed to characterize factors that control the velocity at which groundwater flows between two locations (1 and 2), each of which has a different hydraulic head (h_1 and h_2). Darcy represented the velocity of flow by a quantity called the discharge (Q), meaning the volume of water passing through an imaginary vertical plane perpendicular to the groundwater's flow path in a given time. He found that the discharge depends on the the hydraulic head ($h_1 - h_2$); the area (A) of the imaginary plane through which the groundwater is passing; and a number called the hydraulic conductivity (K). The hydraulic conductivity represents the ease with which a fluid can flow through a material. This, in turn, depends on many factors (such as the viscosity and density of the fluid), but mostly it reflects the permeability of the material. The relationship that Darcy discovered, now known as Darcy's law, can be written in the form of an equation as:

$$Q = KA(h_1 - h_2)/j$$

The equation states that if the hydraulic gradient increases, discharge increases, and that as conductivity increases, discharge increases. Put in simpler terms, the flow rate of groundwater increases as the permeability increases and as the slope of the water table gets steeper.

Groundwater Ecosystems

A groundwater-dependent ecosystem (GDE) is a community of micro-organisms, animals and plants, and associated substrates, whose functioning relies on the presence of water under the ground and its emergence to the surface. Some GDEs are supported entirely by groundwater while others also receive water from different sources, but the groundwater contribution is critical as regards water chemistry to nourish certain species, and provide stable water temperature and absence of sediment load.

GDEs are mainly land-surface features of various types:

- Aquatic – Including springheads, and those wetlands, streams, rivers and lakes receiving groundwater discharge (which are the main focus of attention here).

- Terrestrial – With phreatophyte vegetation, either shallow-rooted in alluvial settings (such as some lowland woods and meadows) or deep-rooted in arid zones with much deeper water-table.

But they can also be subterranean – notably in limestone formations with karstic caverns and fissures inhabited by small invertebrates and some specialised vertebrate species.

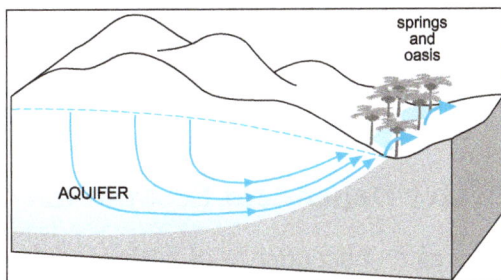

Wetland ecosystem in arid region with only limited contempory groundwater
replenishment and fossil aquifer flow.

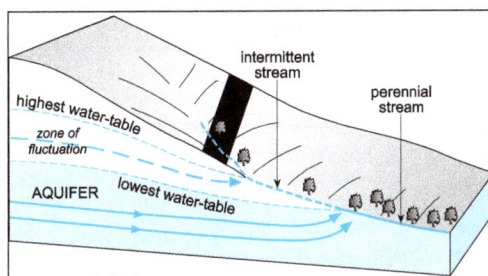

Aquatic streambed ecosystem in humid region along upper reaches of river fed by
perennial and intermittent groundwater discharges.

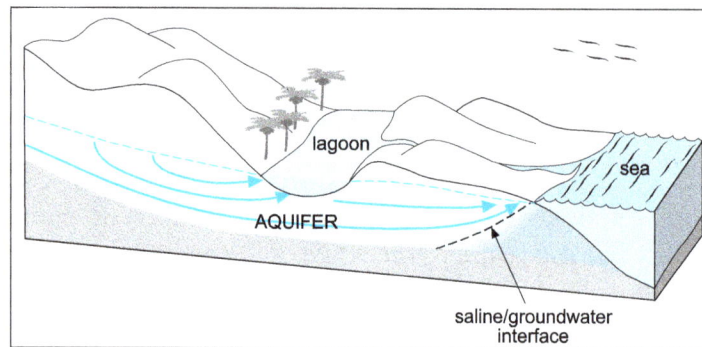

Coastal lagoon ecosystem dependent upon slightly brackish water generated by
mixing of fresh groundwater and limited seawater incursion.

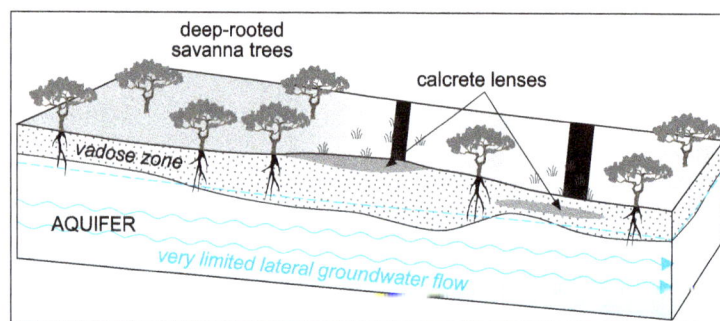

Terrestrial savanna ecosystem dependent upon exceptionally deep-rooted trees
and bushes tapping the water table in arid region.

GDEs are fundamental to the conservation of biodiversity – many being vital for survival of a wide
variety of species. Moreover, GDEs can be of significance as a renewable source of human nutrition
and as key features in the local landscape such as springs and lagoons.

Some prefer to introduce the term 'aquiferdependent ecosystem' to emphasise that aquifers con-
strain the groundwater flow on which ecosystems depend. It should also be noted that some aquatic ecosystems occur in aquifer recharge zones and represent a major source of groundwater system
replenishment.

Degradation of Groundwater Dependent Ecosystems

The degradation of GDEs occurs as a consequence of both modification of natural groundwater
flow regimes and salinisation or pollution of groundwater.

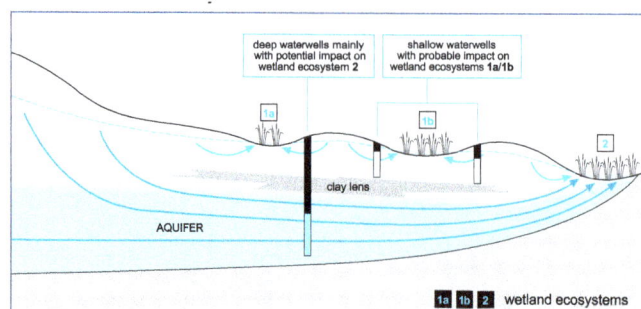

Potential interference with gdes of groundwater pumping from a multi-layered aquifer.

All groundwater pumping has some effect on water-table levels – but in terms of ecosystem impact the main concern is where the cumulative effect of major extraction (for agricultural irrigation or urban water-supply) causes substantial and persistent water-table lowering. This can also sometimes arise in the absence of intensive groundwater use:

- If recharge zones experience major changes that increase consumptive water-use (e.g : lowland afforestation) or riverbed modifications that decrease aquifer replenishment (e.g : impermeabilisation or diversion).

- Where dewatering for tunnel construction or mining operations seriously impacts the groundwater flow regime.

The resulting reduction in groundwater discharge can trigger a change in ecosystem functioning and in turn ecosystem structure, and in extreme cases cause its complete elimination.

There is often uncertainty about the reaction of individual species to hydrological change and species interdependence within a GDE. Moreover, the way that groundwater discharges into, and interacts with, the surface environment (in the transitional hyporheic zone) can be critical to aquatic life (not just the presence of water). Whilst certain irreversible changes can occur in relatively short periods (such as oxidation of wetland or streambed sediments), some species are well adapted to survive hydrologic extremes, and those naturally exposed to variations in groundwater behaviour (for example in ephemeral streams) are more resilient.

Modest increases in groundwater salinity can change ecosystem structure dramatically, and cause extermination of some species. This can occur where irrigation waterwells mobilise salts from depth in aquifers (or contained in aquitards), which are then further concentrated in irrigated soils before leaching to shallow groundwater.

Groundwater pollution can have a similar impact – and the most widespread cause is agrochemical leaching from land-use practices. Pesticides and nutrients that are leached from soils eventually reach the water-table, and can migrate to GDEs in natural discharge zones. The effect of groundwater quality deterioration, notably increases of nitrate, ammonium or phosphate (even at low concentrations) and trace pesticide contamination, may lead to greater ecosystem impacts than groundwater flow diminution. Elevated nitrate and phosphate concentrations can cause ecosystem eutrophication, eliminating oxygen and killing fish and small aquatic animals.

Protection of Groundwater Ecosystems

Two main lines of action are required for protection of GDEs and have been strongly advocated for all sites covered by the RAMSAR Convention:

- Increasing knowledge of their hydrogeological and ecological condition, and the economic consequences of degradation for human well-being.

- Integrating their protection into basin/aquifer water-resource and land-use management.

Systematic assessment is required for each GDE to understand its evolving relationship with underlying aquifers, to evaluate groundwater quality and define chemical baseline, to identify anthropogenic pressure trends, and to determine the socioeconomic contribution of ecosystem services.

The end product should be a conceptual model of GDE functioning, using GIS and other data management systems to display results clearly.

All measures that strengthen the governance and practical management of groundwater can contribute, or be adapted, to the cause of protecting GDEs – by including criteria to maintain groundwater levels and conserve groundwater quality to meet the requirements of the ecosystem receptor. This will often imply greater constraints on the volume and distribution of groundwater withdrawals than would otherwise have been necessary, together with more severe controls over groundwater contaminant load than needed to conserve drinking-water quality. A balance between improving rural livelihoods and sustaining ecosystem health needs to be achieved.

However, social development pressures may be such as to prevent control of groundwater levels and quality in an entire aquifer system in the interest of GDE conservation, and alternatives may need to be pursued such as:

- Acting selectively to introduce 'protection zones' around GDEs, capable of assuring shallow groundwater quality and reducing groundwater level interference.

- Artificial recharge to supplement groundwater flows and improve quality over limited areas in the interest of GDE conservation, or even pumped compensation flows when aquifer levels fall below some critical level.

Value Evaluation of Groundwater Ecosystems

GDE functions are an important component of overall environmental services provided by a groundwater system. Economic assessment of GDEs will require clear definition of the benefits of these ecosystem services including:

- Direct values to the human population in terms of fish and plant production, and providing water storage and purification.

- Indirect values from sustaining biodiversity, habitat and landscape for social, cultural, aesthetic and ethical reasons.

But in the developing nation context it is recognised that direct use values from harvested plants and animals, and from landscapes or habitats for tourism are likely to be more important. If the livelihoods of poor and vulnerable people depend directly on the functioning of GDE's there is a risk that ecosystem value may be systematically under-estimated.

Disputes are often likely to arise over the balance between improving rural livelihoods and sustaining healthy ecosystems. Decision-making needs to be informed by sound technical and economic analysis, and it is thus important to incorporate economic evaluation into the analysis. A useful framework for GDE valuation, and the relationship between human wellbeing and ecosystems, is the so-called 'ecosystem services approach' of the UN Millennium Ecosystems Assessment Project. It argues that human progress has always relied on natural resources and functioning ecosystems. The assessment must use reliable historic data and scientific studies. The method also includes assessment of the impact (observed and forecast) of a set of 'driver factors' that change GDE behaviour.

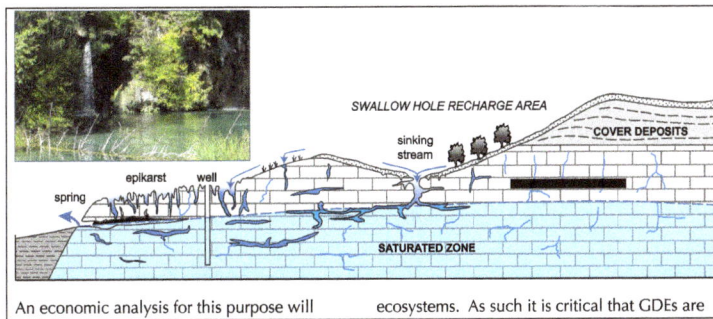

An economic analysis for this purpose will   ecosystems. As such it is critical that GDEs are

An economic analysis for this purpose will normally entail a relative assessment of the:

- Cost of protection in terms of the loss of alternative uses of groundwater and land, and the administration of the land-use and groundwater control policy.

- Benefits of protection in terms of in-situ value of groundwater and groundwater-related ecosystem services.

In this context an analysis based on marginal cost-benefit is likely to better reflect the situation on-the-ground, and scenarios of partial protection (as opposed to total protection) of GDEs will also need to be considered.

Aquifer

An aquifer is a large underground storage space for water. They can be located right at the ground surface or very deep underground where they are impossible to access. They can be very large (e.g. some aquifers can span a province) or quite small.

Aquifers can have high permeability, so water can flow easily in this layer. There is always a barrier below an aquifer that will not let water pass through to lower layers. This barrier, called an aquitard, is a layer with very low permeability. It is very difficult for water to pass through an aquitard, so it helps contain the water in the aquifer. Every aquifer has an aquitard below it and many also have an aquitard above. When an aquifer has an aquitard on top of it, it is called a confined aquifer. If an aquifer only has an aquitard below it and does not have an aquitard above then it is called an unconfined aquifer. Unconfined and confined aquifers are shown in figure.

Unconfined aquifers are directly connected to the surface. When water infiltrates into the ground it passes through the unsaturated zone. The unsaturated zone has mostly air in it and is highly permeable. When the water passes the unsaturated zone it reaches the saturated zone. The saturated zone is filled with water. The boundary between the saturated zone and the unsaturated zone is called the water table. When water seeps into the earth it will travel down the soil layers until it reaches the water table. The water table moves up and down depending on how much water is in the unconfined aquifer. Since unconfined aquifers are connected to the surface, they recharge quickly, and are prone to contamination.

Confined aquifers are confined because they have an aquitard above and below the aquifer. These aquifers do not recharge quickly because it takes a long time for water to pass through the top aquitard. In some cases, confined aquifers contain high quality water because they are not directly impacted by human activity on the surface. Confined aquifers can contain groundwater that is very old. Water can stay in a confined aquifer for several millennia.

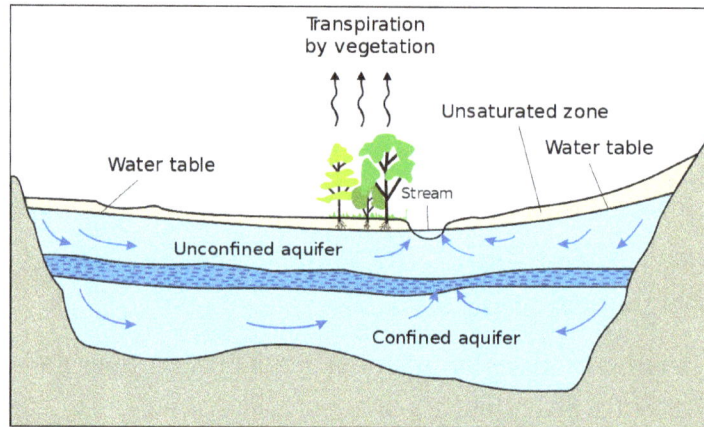

Confined and unconfined aquifers: This figure shows the two types of aquifers, confined and unconfined, within an impermeable bedrock basin. The dark blue layer between the aquifers is an aquitard.

Types of Aquifers

Aquifers are categorized as confined or unconfined, but there are many types of aquifers that are classified by where they are located in the earth and the material of which they are comprised. There are three types of aquifers: unconsolidated deposit aquifers, bedrock aquifers and quaternary aquifers.

Unconsolidated Deposit Aquifers

An unconsolidated deposit aquifer is an aquifer that is made up of loose sediment such as gravel and sand. These aquifers are close to the surface and are almost always unconfined. This type of aquifer is commonly found near rivers in a floodplain. Unconsolidated deposit aquifers are formed as the result of old rivers that no longer exist, by glaciers that have moved the sediment or by deposition at the bottom of a lake. The water in an unconsolidated deposit aquifer is directly connected to the surface water system.

Bedrock Aquifers

Bedrock is the hard rock that lies below all the sand, gravel and soil near the ground surface. A bedrock aquifer is an aquifer that is confined within hard bedrock layers. Water can travel through porous bedrock, or through cracks, fractures and crevasses in the hard bedrock. In Alberta, 84% of groundwater wells draw from bedrock aquifers. These aquifers are easily accessible in areas where the bedrock is near the earth's surface, such as in southern Alberta.

In Alberta, there are three types of bedrock aquifers: carbonate aquifers, sandstone aquifers, and fractured shale aquifers.

Carbonate aquifers are made of rocks such as limestone and usually contain saline water. Sandstone aquifers are made of sandstone, a highly permeable rock, and can contain either saline or fresh-water. The largest aquifer in Alberta, the Paskapoo Aquifer, is a sandstone aquifer. One third of groundwater wells in Alberta are located in the Paskapoo Aquifer. Shale is a rock that is similar to sandstone, but is less permeable. For shale to be an aquifer, it must be fractured, or cracked, so water can flow into it. Fractured shale aquifers are relatively rare in Alberta. The wells that draw from this type of aquifer do not produce as much water.

Quaternary Aquifers

Quaternary aquifers are aquifers that were created by glaciers. They are located between bedrock and the earth's surface. These aquifers can be confined or unconfined. There are two types of qua-ternary aquifers: buried valley aquifers and alluvial aquifers.

Buried valley aquifers are confined aquifers that can be directly above bedrock or higher up in the rock layers. These are ancient valleys that are filled with permeable sand and gravel. Unconfined sand and gravel aquifers are located at the surface or near the surface. An alluvial aquifer is a spe-cific type of unconfined aquifer which has a river flowing through it. The river is the main source of recharge. Quaternary aquifers generally contain freshwater.

Water Movement in Aquifers

Water movement in aquifers is highly dependent of the permeability of the aquifer material. Per-meable material contains interconnected cracks or spaces that are both numerous enough and large enough to allow water to move freely. In some permeable materials groundwater may move several meters in a day; in other places, it moves only a few centimeters in a century. Groundwater moves very slowly through relatively impermeable materials such as clay and shale.

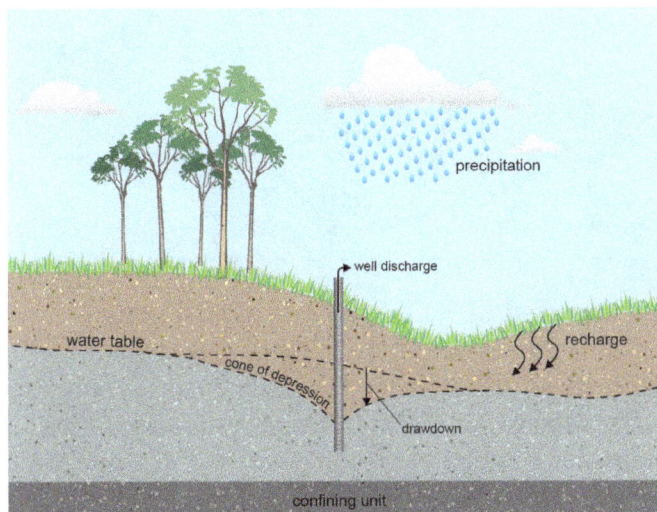

Schematic showing a cone of depression around the well,
usually the result of over pumping.

After entering an aquifer, water moves slowly toward lower lying places and eventually is dis-charged from the aquifer from springs, seeps into streams, or is withdrawn from the ground by wells. Groundwater in aquifers between layers of poorly permeable rock, such as clay or shale, may be confined under pressure. If such a confined aquifer is tapped by a well, water will rise above the top of the aquifer and may even flow from the well onto the land surface. Water confined in this way is said to be under artesian pressure, and the aquifer is called an artesian aquifer.

The Hydrologic Cycle

The Hydrologic Cycle is one of the most important processes in the natural world, and is perhaps one that we all take for granted. All of the world's water is subject to this process, which sees the water change forms, locations, and accessibility.

In its most basic assessment, water changes between three different states in this cycle. It variously takes the form of liquid, gas, and solid: water, steam or vapor, and ice. Throughout the cycle, the water will undergo changes between these three forms many times: water freezing into ice, ice melting into water, water evaporating into water vapor, and that vapor then condensing to become water once more.

The hydrologic cycle, also known as global water cycle or the H_2O cycle, describes the storage and movement of water between the biosphere, atmosphere, lithosphere, and the hydrosphere.

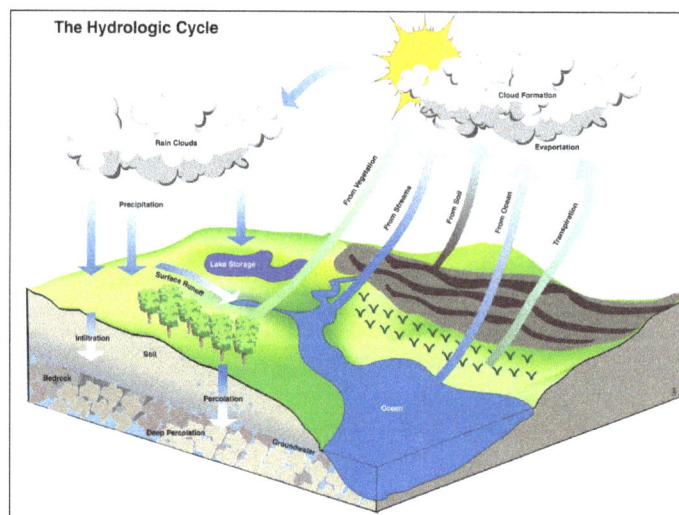

Water is most commonly found in its liquid form, in rivers, oceans, streams, and in the earth. The sun's rays constantly warm the water found in these places and, whether through this heat or through man-made means, the water particles gain energy and spread, turning the water from a liquid into a vapor through evaporation. The water vapor, thus becoming less dense, rises with the warm air into the sky where it sticks to other water particles to form clouds.

Typically, we consider the boiling point of water to be a hundred degrees centigrade, which is certainly true when pressure and humidity are normal. However, places such as mountains, where humidity is low and pressure is even lower, require less energy to boil away the water.

Along with the water vapor, some small particles can often rise up to form clouds. It is not only liquid water that can evaporate to become water vapor, but ice and snow, too. This process is simple enough, however there are a few things to note about evaporation.

This simple explanation, however, does not do justice to the complexity of the hydrologic cycle, which comprises many more steps. Here is a breakdown of the different steps of the hydrologic cycle.

Different Steps of the Hydrologic Cycle

Here is a breakdown of the different steps of the hydrologic cycle.

Water is most commonly found in its liquid form, in rivers, oceans, streams, and in the earth. The sun's rays constantly warm the water found in these places and, whether through this heat or through man-made means, the water particles gain energy and spread, turning the water from a liquid into a vapor through evaporation. The water vapor, thus becoming less dense, rises with the warm air into the sky where it sticks to other water particles to form clouds.

- Evaporation – Is frequently used as a catch-all term to refer to the process of water turning to water vapor, however there is another distinct term for the evaporation of water from a plant's leaves.

- Evapotranspiration – Makes up a large portion of the water in the planet's atmosphere due to the sheer surface area of the globe covered by flora. The majority of water in the atmosphere comes from lakes and oceans – around ninety per cent – but in terms of land-based water, evapotranspiration is an important player.

- Sublimation – As the process is called, results from when pressure and humidity are low as noted above. It is not only liquid water that can evaporate to become water vapor, but ice and snow, too. Due to lower air pressure, less energy is required to sublimate the ice into vapor. Other factors which can aid in sublimation are high winds and strong sunlight, which is why mountain ice is a prime candidate for sublimation, while ground ice sublimation is not so common. A good, visible example of sublimation is dry ice, which emits a thick layer of water vapor due to its lower energy requirement.

The further above sea level one gets, the cooler the air. When water vapor reaches this plane, it cools significantly and clumps together. So stuck together, this newly formed cloud is subject to the movement of the wind and the changes in the air pressure, which is what moves the water around the planet. There are a couple of things that can happen to the vapor in this state.

- Precipitation/Rainfall – Refers to vapor that cools to any temperature above freezing point (zero degrees centigrade) will condense, becoming droplets of liquid water. These droplets form when the water vapor condenses around particles and other matter that rises up with the water during evaporation, giving a nucleus to the water droplet so that it can clump together. Once a number of this tiny, particle based droplets form, they collide and clump together as larger droplets. At a certain point, the droplet will become big enough that its mass will be subject to the force of gravity at a rate faster than the force of the updraft in the air around it. At this point, the water falls to earth.

- Snow – Refers to frozen water falling from the sky. When it is particularly cold or the air pressure is exceptionally low, these water droplets will crystalize before falling.

- Sleet – Is a bitterly cold, half-frozen slush. This third state occurs when the conditions are not quite cold enough to keep the crystals frozen and the water either does not freeze fully or if precipitation occurs in particularly cold conditions, or conditions in which the air pressure is very low, then these water droplets can quite often crystallize and freeze. This causes the water to fall as solid ice, known melts somewhat in the process.

When water falls to earth, it quite often ends up on tarmac or over man-made surfaces where it quickly evaporates again.

- Infiltration – Is water that doesn't evaporate after precipitation and falls into soil and other absorbent surfaces. The water moves throughout the soil, saturating it.

- Groundwater Storage – Is water that has not precipitated or run off into streams or rivers, but instead moves deep underground forming pools known as "groundwater storage". In groundwater storage, water joins up in the soil and forms pools of saturated soil instead of escaping the soil. These pools are called "aquifers".

- Springs – Occur when an aquifer becomes oversaturated, and the excess water leaks out of the soil onto the surface. Most commonly, springs will emerge from cracks in rocks and holes in the ground. Sometimes, if conditions are particularly volcanic, the spring will heat up and form "hot springs".

- Runoff – After heavy rainfall has saturated the soil it will cease to absorb water and additional rainfall, as well as melted snow and ice, will simply flow off of the surface. The flow follows gravity down hills, mountains, and other inclines to form streams and join rivers. This is known as "runoff", and it is the principle way in which water moves along the Earth's surface. The rivers and streams are pulled by gravity until they pool together to form lakes and oceans.

- Stream flow – Is the direction the runoff takes to form a stream and it is this flow which dictates the river's currents depending on how close they are to the ocean. Because ice and snow make up a large portion of the water involved in runoff, heat waves are a principle cause of flooding as the water stored on the surface is suddenly released into runoff flow. In particular, a warm spring following a cold winter can result in quite spectacular flood, as a large volume of water gets stored in ice and snow only to quickly melt and form new streams.

- Ice Caps – Occur when a large volume of snow falls and is not evaporated or sublimated, the ice compacts under its own weight to form these caps. Ice caps, glaciers, and ice sheets contain a huge amount of water, and those found in the polar regions of the planet are the largest stores of ice found in the world. As the atmosphere warms up slowly, more and more of this ice melts and evaporates, releasing more water into the hydrologic cycle. It is this process which causes rises in the ocean levels.

The hydrologic cycle happens continuously, with all different steps happening simultaneously around the world. The biggest concern that many have with the hydrologic cycle is the availability of drinkable water, which is something that is constantly in flux, and the melting of the huge ice storage sheets at the polar caps. Having an understanding of the different steps of the hydrologic cycle is an important step in understanding what effect human activity has on the world's water.

References

- Hydrogeology, geology-and-oceanography, earth-and-environment: encyclopedia.com, Retrieved 2 June, 2019

- Groundwater: newworldencyclopedia.org, Retrieved 14 January, 2019

- IAH-SOS-Ecosystem-Conservation-Groundwater: iah.org, Retrieved 25 April, 2019

- What-is-an-aquifer: albertawater.com, Retrieved 5 March, 2019

- Science-center-objects, aquifers-and-groundwater, science, water-science: usgs.gov, Retrieved 26 August, 2019

- Different-steps-of-the-hydrologic-cycle: conserve-energy-future.com, Retrieved 6 February, 2019

Permissions

We would like to thank the editorial team for lending their expertise to make the book truly unique. They have played a crucial role in the development of this book. Without their invaluable contributions this book wouldn't have been possible. They have made vital efforts to compile up to date information on the varied aspects of this subject to make this book a valuable addition to the collection of many professionals and students.

This book was conceptualized with the vision of imparting up-to-date and integrated information in this field. To ensure the same, a matchless editorial board was set up. Every individual on the board went through rigorous rounds of assessment to prove their worth. After which they invested a large part of their time researching and compiling the most relevant data for our readers.

The editorial board has been involved in producing this book since its inception. They have spent rigorous hours researching and exploring the diverse topics which have resulted in the successful publishing of this book. They have passed on their knowledge of decades through this book. To expedite this challenging task, the publisher supported the team at every step. A small team of assistant editors was also appointed to further simplify the editing procedure and attain best results for the readers.

Apart from the editorial board, the designing team has also invested a significant amount of their time in understanding the subject and creating the most relevant covers. They scrutinized every image to scout for the most suitable representation of the subject and create an appropriate cover for the book.

The publishing team has been an ardent support to the editorial, designing and production team. Their endless efforts to recruit the best for this project, has resulted in the accomplishment of this book. They are a veteran in the field of academics and their pool of knowledge is as vast as their experience in printing. Their expertise and guidance has proved useful at every step. Their uncompromising quality standards have made this book an exceptional effort. Their encouragement from time to time has been an inspiration for everyone.

The publisher and the editorial board hope that this book will prove to be a valuable piece of knowledge for students, practitioners and scholars across the globe.

Index

www.ingramcontent.com/pod-product-compliance
Lightning Source LLC
Chambersburg PA
CBHW061246190326
41458CB00011B/3597

9 781641 165808